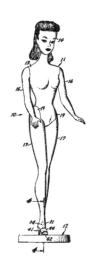

Inventing the American Dream

A HISTORY OF
Curious, Extraordinary, and
Just Plain Useful Patents

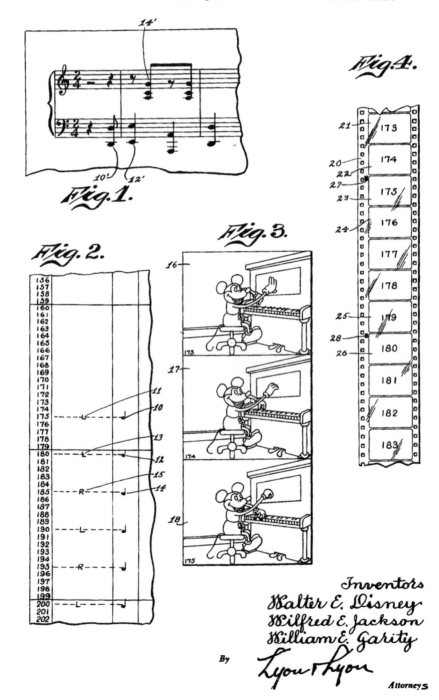

Inventors
Walter E. Disney
Wilfred E. Jackson
William E. Garity

By

Lyon & Lyon

Attorneys

Inventing the American Dream

....

A HISTORY OF

Curious, Extraordinary, and Just Plain Useful Patents

Stephen van Dulken

THE BRITISH LIBRARY

First published 2004 by
The British Library
96 Euston Road
London NW1 2DB

British Library Cataloguing in Publication Data
A CIP record is available from the British Library

ISBN 0-7123-0893-8

Designed by Bob Elliott
Typeset by Hope Services (Abingdon) Ltd.
Printed in England by St Edmundsbury Press,
Bury St Edmunds

All measurements in this book are given in American units.
Below is an approximate conversion table to metric measurements.

1 mile	=	1.6 kilometres
1 yard	=	0.9 metre
1 foot	=	0.3 metre
1 inch	=	2.5 centimetres
1 acre	=	0.4 hectare
1 gallon	=	3.8 litres
1 ton	=	1 tonne
1 pound	=	0.5 kilogramme
1 ounce	=	28 grammes
1 horsepower	=	750 watts
1 °F	=	$1.8 \,°C + 32$

Contents

US State abbreviations

Localities mentioned in this book have their state given as standard postal abbreviations (unless they are well-known cities). All the American states are listed below, although some are not mentioned in the book.

State	Abbr.	State	Abbr.
Alabama	AL	Montana	MT
Alaska	AK	Nebraska	NB
Arizona	AZ	Nevada	NV
Arkansas	AR	New Hampshire	NH
California	CA	New Jersey	NJ
Colorado	CO	New Mexico	NM
Connecticut	CT	New York	NY
Delaware	DE	North Carolina	NC
Florida	FL	North Dakota	ND
Georgia	GA	Ohio	OH
Hawaii	HI	Oklahoma	OK
Idaho	ID	Oregon	OR
Illinois	IL	Pennsylvania	PA
Indiana	IN	Rhode Island	RI
Iowa	IA	South Carolina	SC
Kansas	KS	South Dakota	SD
Kentucky	KY	Tennessee	TN
Louisiana	LA	Texas	TX
Maine	ME	Utah	UT
Maryland	MD	Vermont	VT
Massachusetts	MA	Virginia	VA
Michigan	MI	Washington	WA
Minnesota	MN	West Virginia	WV
Mississippi	MS	Wisconsin	WI
Missouri	MO	Wyoming	WY

Preface

● ●

THIS book looks at the huge and exuberant world of patents and how they have helped to create the American Dream, even if that Dream of ease, comfort and prosperity always seems to many to be at least a little out of sight. It is illustrated throughout by drawings from patents with a few from trademarks. It is an unusual angle from which to examine America, but a rewarding one, as I hope to prove. Innovations do not come out of thin air: everything has to be thought of, developed, and marketed, and not all good ideas or aspirations succeed. As patents mostly reflect material things, or ways of making them, and hope in inventors' hearts always springs eternal, they are particularly appropriate. I have tried hard to squeeze as much as possible in, but as there is so much material I have inevitably had to be very selective, and I have tried to avoid repeating material in my earlier books. Incidentally, not all those who contributed to the Dream were Americans.

To better understand this book here is a *brief* explanation of the world of intellectual property, and especially patents, as it is experienced in the USA.

The patent system originates from the Constitution, where Article 1, section 8 states: "To promote the Progress of Science and useful Arts, by securing for limited Times to Authors and Inventors the exclusive Right to their respective Writings and Discoveries". This has always meant that the inventor, or *patentee*, is the person who is emphasized. If the rights in the patent is made over to, say, a company or a university then they are the *assignees*. A patent must be new or at least non-obvious to those skilled in the art. This novelty is measured not by the date the patent was filed (as now practised everywhere else in the world) but rather by who invented it first. An *interference* is a dispute about who has prior rights which is heard by Patent Office officials. Protection was for 14 years after which it was free for anyone to use. The original system where the President and members of the Cabinet were involved was soon dropped. The archives were saved in the War of 1812 from being burned when the head of the Patent Office made an impassioned plea to British army officers, only for all 10,000 patents to get burned in 1836. Subsequent attempts to get copies from the inventors meant that about a quarter were replaced, but most were lost for good, although some exist as English or French patents where the inventors applied for protection in those countries.

Also in 1836, patents were examined by patent office examiners to see if they were new, and the patents were numbered for the first time from number 1. It took until 1911 to reach 1 million, while the 6 million mark was reached in 1999. This is despite the confident assertions on many a website that the Commissioner for Patents for 1898–1901, Charles Duell, said in an annual report to Congress that "everything that can be invented has been invented". Strangely enough, an inspection of the reports does not reveal such a statement. In 1842 a system of *design patents* was started, which protect the appearance of an object. These too were numbered from number 1 and over 470,000 have been granted since then. To distinguish them from the other kind, *utility patents* is used for those patents which are for function. This book uses "patents" to mean the main kind, in the format US 5000000, and US D20000 for the design patents. In 1861 the patent term was extended to 17 years from the date of grant (or publication, which was the same date).

A national *trademark* scheme was started in 1870, although state trademarks also exist. These were at first words or pictures or a mixture which depicted the goods for sale although they can now be, for example, sounds or smells. All these forms (plus labels on packaging etc.) were recorded in the *Official Gazette*, which was printed from 1872, although design patents were not depicted until 1892. *Plant patents*, a unique concept, were introduced in 1930. In 1980 the idea of paying maintenance fees to keep the patent going was introduced for the first time, but unlike many countries there is no requirement to "work" the invention. In 1995 the length of terms changed to 20 years from date of application. This prevented "submarine" patents which emerged after many years to take advantage of a term following grant. The most notorious of these was George Selden's US 549160, which was applied for in 1879 and was granted in 1895. This patent for a "Road engine" gave him the monopoly over the car industry even though his model was hardly capable of moving. Many people have lost their jobs when assembly lines have abruptly had to stop work owing to the emergence of submarine patents.

Also in 1995 there was the provisional application, which gave a year's grace in filing for purposes of novelty. This was useful when applying abroad as this must be done within 12 months of filing at the national patent office, and an extra year is hence available to inventors. Probably the biggest shakeup in the patent system was when applications made from 29 November 2000 had to be published as applications 18 months from the original filing date. This follows the practice of nearly every industrialized country. This helped again with submarine patents, as those watching for patents in their field can monitor what is happening at an early date and can make objections. These applications are numbered in the format 2001/0000001. There was an exception made for those who wished only to file in the USA (and not in a foreign country) who do not need to publish at this preliminary stage. This is unique in the world, and resulted from pressure from private inventors who believed that early publication would mean that their inventions would be stolen.

In addition to such changes, the scope (as well as number) of utility patents has widened considerably in recent decades. Patents for software, business methods and microorganisms among others have emerged as there is pressure from industry for more protection. Copyright is available as a right which can be registered at the Library of Congress. At present this is for the lifetime of the creator plus 70 years, and was formerly used for software (as well as books, movies and so on). However, as it is a right against deliberate copying it is a weaker protection than a utility patent. More changes will no doubt occur in the future, as so much money is often dependent on patent rights.

I would like to express my gratitude to my patient editor, Anthony Warshaw and the designer, Bob Elliott.

Hush-a-bye, baby, in the tree top

C HILDHOOD is a precious thing, and a happy upbringing is surely essential. In the old days this might have consisted of splashing about in swimming holes on a hot day, cycling around the neighborhood, and cool drinks of lemonade. Nowadays things are different. Children seem to be forced to grow up quickly, if only by passively watching advertisements on television which encourage them to demand yet more toys, gadgets or clothes to boost their self-esteem and to avoid feeling left out when with friends. Surely all that watching (and not just of commercials) is having an effect, otherwise the advertising agencies are wasting a lot of their clients' money. It is certainly a lucrative market, estimated at over $21 billion being spent annually on (or by) children aged 14 and under. Children also *want* to grow up, with a typical reply to a query about their age being "Seven. Nearly".

There is more activity, perhaps, by private inventors in the field of devices for children than in any other area. Over 8,000 American patents since 1920 have had words like "child" or "baby" in the title, and many inventions for the market of course do not use such words. People patent what they care about, and where they see a need, and parents as well as professionals working with children are obviously the most likely to notice where improvements can be made. It is therefore almost certainly the field with the highest number of women private inventors.

Starting at the beginning, there have been over 120 American patents in class 436/65, which is for methods of determining if pregnancy or ovulation has occurred. Some are no doubt for private use, but the earliest in that class was intended "to be used by a physician with very little equipment other than that usually found in a physician's office". This was US 2564247, which was applied for in 1947 by Richard Carson and Reuben Sacks of Dayton, OH. The title was "Method of testing urine to determine pregnancy". The idea was that histidine, an amino acid, was known to be normally metabolized in the liver into histamine. In pregnant women, however, histidine was excreted instead. The "specimen" had to be diluted in water and chemicals such as a pH indicator were added and the mixture heated. A scale would then indicate if a significant color change had occurred. Nowadays tests would normally check for human chorionic gonadotropin in the sample.

Besides finding out if the happy event has occurred, it is possible to provide the unborn baby with the American Dream *before* birth. Kevin Harrison of Montgomery, AL, filed in 1997 for US 5873736, a "prenatal music belt". There is much evidence that babies in the womb do respond to outside stimuli such as music, so this is an interesting idea. The patent adapted the concept used in US 4798539 by Verlyn and Danise Henry of Lakeview, MI, which dates from 1987. Harrison pointed out problems with the Henry idea such as its being heavy and uncomfortable, and the inability to accommodate those of different sizes with its Velcro® straps, including the problem of girth growth during pregnancy. Needless to say, Harrison felt that he had solved these problems with an adjustable belt which had pockets in which the earphones were placed.

For childbirth itself, Charlotte and George Blonsky of New York City filed in 1963 for their US 3216423, entitled "Apparatus for facilitating the birth of a child by centrifugal force". This patent is famous among those working in the intellectual property field. It does exactly what it says, as the patent explains that "primitive peoples" carry on exercising during pregnancy so that "nature provides all the necessary equipment and power to have a normal and quick delivery". Not so, it says, the American, who often did not have the same opportunities of manual labor. The patent explained, in nine pages, how spinning the mother in a controlled manner would spin the baby out with the aid of her own efforts. A handbrake enabled the "operator" to switch the machine off once birth had occurred, or on instructions from the "supervising gynaecologist". This invention may sound very odd, and probably would not have (safely) worked if ever tried out, but novelty and not the ability to work is what is required of a patent. The patent itself is indexed as subclass 606/121, "Parturition assistance device", which has some 50 others in the same subclass, at least one being designed to help hogs. After the birth, although the mother has done all the hard work the proud father is (at least in theory) supposed to pass cigars around. Joseph Swarbrick of North Arlington, NJ, filed in 1949 for US D160562, a "reversible cigar band", to allow for either eventuality.

Besides numerous inventions for babies such as rattles, pull-toys and so on, designed to entertain and perhaps educate the growing child, there are more practical devices such as pacifiers. These date back at least to an 1860 British patent, and there are over 100 American patents in subclass 606/235, "Oral pacifier: teething device", while neighbor 606/235, "Oral pacifier: nipple attachment or structure", has over 160. These include a tiny pacifier designed for low-weight babies. This invention was filed for in 1991 by three women staff at the School of Nursing of the University of Texas-Houston Health Science Center. US 5275619 explains that babies in the womb like thumb-sucking, and the miniature pacifier is designed to simulate the feeling. Such babies are often too small to get their thumb in their mouth and hence soothe themselves. The pacifier also means that the baby gets used to the sucking reflex so vital to feeding. Dr Joan Engebretson, one of the inventors, said in a 1994 press release that "We're hoping our pacifier will help these babies get ready to be fed on the breast or on the bottle", especially as trials showed that babies using the pacifier remained in an alert state for longer periods

Reversible cigar band (US D160563)

of time than those not using the pacifier, and, as she said, "Alert babies feed better than drowsy babies".

Thumb-sucking itself has been the subject of dozens of hostile patents. These generally involve glove-like structures but there are exceptions. One is the "Thumb-sucking alarm system", US 4178589, dating from 1978. It consists of a "pair of spaced-apart sensor electrodes carried on a moisture-accepting portion of a flexible sensor tape" wrapped round the thumb. A test button was provided to check that the system was working properly. Another is US 3552805, dating from 1968, where a device attached to the upper teeth made the unfortunate toddler look like Dracula.

To keep them out of mischief babies can be placed in a crib. There are over 550 American utility patents for the basic idea of a crib—apart from subdivisions to allow for concepts such as self-rocking, swinging, folding or dismountable cribs (some doubling as walkers), and even a "hydrostatic steam cradle". Both it (from 1827) and the earliest American patent for a crib, from 1815 by S. Pope of Windsor, CT, were sadly destroyed in the 1836 fire at the Patent Office, and it was not possible at the time to find replacement copies of how the inventions worked. The self-rocking cradles do sound risky if the baby has just been fed.

A hazard with cribs is that the baby may move around in such a way that the fitted sheets come loose, and the child can get entangled and suffocate. This happened to Marie Reen of Norwood, MA, when she lost her 13-month-old son Jimmy. He was wrapped up so many times that only his feet were visible. After his death she wanted to prevent the tragedy happening to others. She found that since 1984 at least two other deaths had occurred from the same cause and that the sheets were about the only aspect of the crib which did not have to meet Federal safety standards. For example the mattresses have had to conform to size standards since 1973, but not the sheets. The Good Housekeeping Institute was inspired by Reen to try out 22 brands of sheets. Each was washed five times and they were surprised at the difference that made in the ability to fit on the mattress. They also tried using a force-gauge on a 20-lb sandbag to simulate a one-year old's force on the sheets. Only four sheets were considered adequate after the testing.

She applied in 1998 with her mother Charlene Reen for what became US 6067677. The StayPut™ sheet fits on the mattress like a pillowcase. It slides onto the mattress through an opening in one end of the sheet. The perimeter of the opening has a continuous elastic band which allows the sheet to fit tightly over the mattress without any snagging. The sheet is secured by a flap, attached to the open end of the sheet, which is pulled over the encasement opening and is secured by hooks and loops. The product, sold by their own company JR Safety Crib Sheets, has had a lot of favorable publicity. About 40 children die annually from injuries associated with cribs. "This is a decrease from the estimated 150 to 200 deaths in the 1970s", states the Consumer Product Safety Commission, but grieving parents know that even one death is too many. Another cause of deaths in young children is SIDS, or Sudden Infant Death Syndrome. Following a campaign to reduce the numbers who die from this tragic condition by encouraging the laying of babies on their backs, deaths fell by 38% between 1992 and 1995. Even so some 3,000 babies still die annually because of SIDS. There are over 200 patents suggesting methods of monitoring or prevention, typically based on devices sensing the stopping of the heart.

On a more reassuring note, Thomas Zelenka of Hanford, CA, filed in 1968 for US 3552388, a "Baby patting machine". The baby is placed in a crib and a motorized paddle periodically pats the child. The wording is attractive. "It is generally well-known to most parents of small infants and children that it is sometimes difficult for the infant to fall asleep, and the parent must resort to patting the baby to sleep by repeated pats upon the hind parts thereof. This can be a time consuming operation particularly when the infant is restless and not likely to fall asleep easily, and it is particularly objectionable to the parent when this takes place during the night, thereby disturbing the parent's own sleep. This situation is accordingly in want of improvement."

Many a design has no doubt been made or pondered over for clothes that would take rough treatment, but few have caught on. One that did not was Ellen Nester's

US 1424318, filed in 1922 from Gardner, MA. The garment was in one piece and lacked buttons. The child placed the legs into short "bloomers" and then pulled the blouse, which was sewn to the bloomers at the back, over her head and body. Then there was Patricia Grose of Highland Park, MD, in 1999 with US 6076186, a "Crib climbing restraint garment for toddlers". It involved placing children in sack-like garments which, of course, meant that they could not use their legs. It is also expensive to keep buying new clothes every time a child goes up a size. Diana Drmaj of Ontario, Canada suggested with her US 4259751 in 1978 turning up the material at the hem or cuff twice. Besides the permanent stitching there would be easily removable stitching. When cut, the garment would lengthen 1 or 2 inches. This sounds useful for parents but might not have sounded so good for the manufacturers, with fewer sales.

The idea of carrying a baby in a sling or by some other arrangement close to the body has been known for a very long time. The American Indians habitually used a papoose. Despite this example the adoption of such garments seems to have been slow, and US 484065 is an unusual patent for its time. It was by Charles Taylor of New York City and dates from 1892. Page 6 shows an excellent depiction of contemporary smart fashion, but baby looks rather precarious. The idea, Taylor said, was to distribute the weight of the baby round the body.

Nowadays many parents have a similar, if more reliable, baby carrier. Their popularity is thought to date from when Ann Moore was working as a pediatric nurse with the Peace Corps in Togo in the early 1960s. She realized that the slings that African women used to carry their babies provided a great deal of comfort and security, while keeping the parent's hands free. The credit is given to her, but US 3481517 is in the name of Agnes Aukerman (her mother, as it happens), of West Alexandria, OH, who filed for the patent in 1968. The patent mentioned the growing popularity of the concept but stated that frames were awkward and could cramp the baby. It was in the shape of a knapsack with straps below and above the shoulders with baby peering out. The product was trademarked as Snugli®, a rugged, adjustable, pouch-like infant carrier, and (adapted) products are still selling under that name.

There is a long history of means for conveying children in wheeled vehicles. An early and unusual example was US 308467 by George Clark of Dubuque, IA, which dates from 1884. The perambulator is in the shape of a huge shoe with, it seems, four children (or dolls, the patent states) sitting in it. The inventor may have been inspired by the rhyme about the old woman who lived in a shoe. More typically strollers have been around for a long time, with US 2711328, which is from 1952, an early and sophisticated example of one. The patent was by Strollee of California, a Los Angeles company. Page 7 shows a "Folding baby stroller".

The stroller folded by pivoting the handle (132) to release a lock at its base. It then telescoped and pivoted inwards to make a (relatively) "compact bundle" which could be gripped by the top of the handle and carried off. By removing

Fig.1.

Fig.2.

Baby carrier (US 484065)

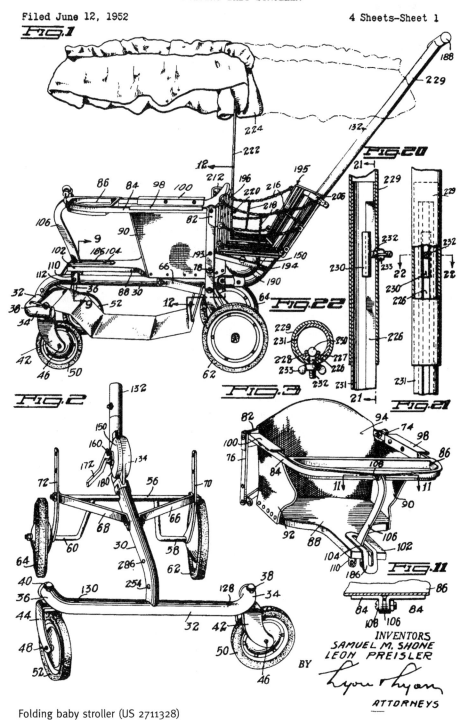

FIG.1

FIG.20

FIG.22

FIG.3

FIG.21

FIG.2

FIG.11

INVENTORS
SAMUEL M. SHONE
LEON PREISLER
BY

ATTORNEYS

Folding baby stroller (US 2711328)

several bolts the seat could become an infant's seat in a car, or alternatively the framework could take a wagon (available separately) for when the child was too old for a stroller.

Wagons have certainly been popular for a long time. The best known child wagon is the Radio Flyer®, which originated with Antonio Pasin, who arrived in 1914 from Italy as a penniless 16 year old. "The little red wagon" has one child sitting in it and steering while a second child pushes it. The family were cabinet-makers, and Pasin tried making wooden wagons which children could tow along. Although they sold well, he realized that anything made of wood was likely to be handmade and hence slow to produce. He looked at the car industry, and adopted from it metal-stamping techniques and other mass-production ideas, so that he came to be nicknamed "The little Ford". The trademark Radio Flyer came from two fairly new ideas at the time (one being Italian in origin): Pasin was simply trying to be trendy.

Pasin filed in 1933 from Chicago for this, his first patent, a "Coaster wagon". It was the Streak-O-Lite model, or is at least very close to it. The patent explains that it was designed to be as simple as possible to manufacture. The Streak-O-Lite was meant to resemble the Zephyr trains, even boasting headlights and control dials. The third generation of Pasins continue to work in the Radio Flyer company, developing variations on the same basic, and much-loved, principle.

Many a despairing parent has tried to think of ways of bribing their children into doing chores or just being "good". Others feel that virtue should be its own reward. Helen Norford of Chicago, IL, with her US 2340139, filed in 1943, inclined towards the latter with her "Indicating climbing device". It was for a sheet with a real ladder attached (but only fit for the large two-dimensional figures of children shown in the drawing, which could be clipped into place). A moralistic poem was printed on the ladder and every achievement during the day was rewarded by moving the figure up to the top, which represented bedtime. It would have been particularly apt if placed against a bunk-bed. The complete poem, which can only be partially read in the drawing, is included in the patent description. It ends with:

> Now I quickly undress and off to bed
> But not to sleep till my prayers are said
> Stars are many Stars are bright
> I have won my star tonight.

The activity was said to be "morale building". Alfred Blaine of St Petersburg, FL's effort in 1955 with his US 2883765 was more like bribery. His "Child's chores recorder for producing incentive" was a complete system which would "provide a clear, running record of the rewards or forfeitures acting as a measure of a child's cooperation in the home or schoolroom". Blaine makes it sound like Boot Camp.

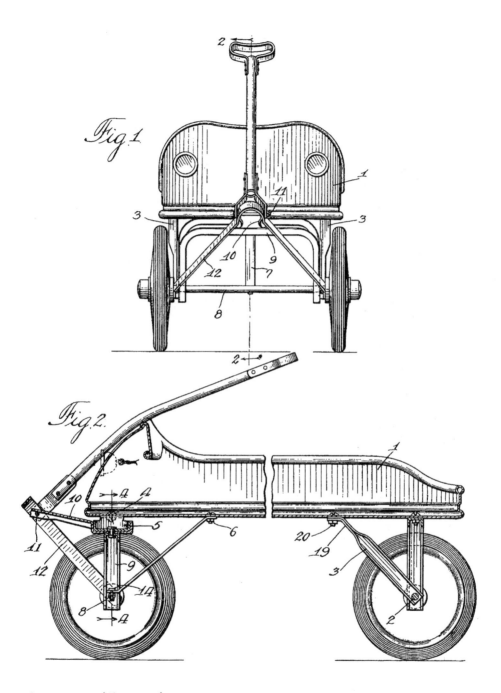

Coaster wagon (US 2015726)

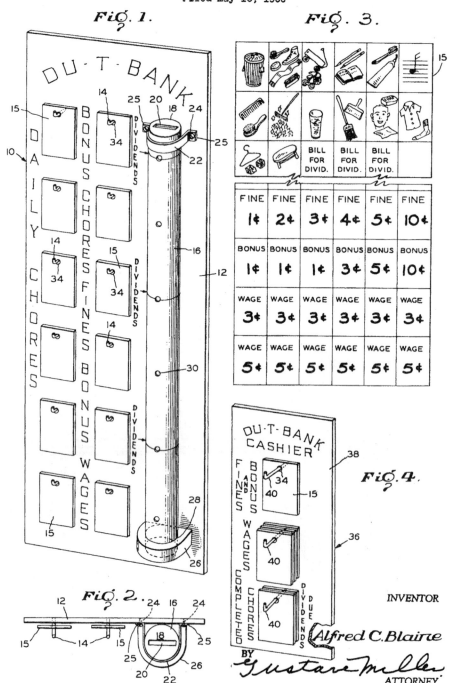

Chores recorder (US 2883765)

Figure 1 is the "assembly", with Figure 2 showing it from above. Figure 3 is meant to be cut up to provide the "checks" which are placed on the assembly, some being required chores and some voluntary, as well as wages and fines. Figure 4 is the "cashier board". The basic idea was that at the beginning of the week the checks corresponding to all wages, bonus money and fines were placed on the top two hooks of the cashier board. As the child completed chores, whether required or bonus, the corresponding check was hung on the bottom hook. At the end of the week the teacher or parent "redeems for cash all wage and bonus checks, less the total amount of the fines". Brushing your teeth seems to have been identified as a chore, while the musical note next to it in Figure 3 was probably meant for playing an instrument rather than a reward for not playing loud music. Even for 1955 the amounts of money involved must have been tiny, while it does not seem to have occurred to Blaine that his system was open both to cheating and to pranks.

Soiled diapers are a constant nuisance for young families. New mothers used to be told to buy "a dozen of everything" since laundering was an exhausting and time-consuming task. A big advance was when in 1949 Marion Donovan of Saugatuck, CT, applied for US 2556800 and later other patents. It took her years to develop the concept. A plastic cover was placed around the traditional cloth diaper and both could be washed and reused. Originally she wrapped shower curtain or parachute fabric (stories differ) round the traditional cloth diaper, and she also was the first to introduce snap-buttons to replace dangerous safety pins. Her invention was called the "Boater" as it prevented wet diapers from leaking (and so sinking the "boat"). In a 1993 interview she said that her invention resulted from frustration with messy baby clothing and bedding, and with diaper rashes. "It was a horrible, arduous job to deal with. I just wanted to contain the wet from going everywhere."

The next step towards the American Dream was of course the fully disposable diaper, which today has over 90% of the market. Its invention is attributed to Procter & Gamble with a paper-based version. Being paper, of course, it could hardly be reused. Procter & Gamble had acquired in 1956 a paper pulp plant, and Victor Mills had the job of figuring out what to do with the plant. Mills, a grandfather, was reminded of how much he had hated changing diapers. It occurred to him that the pulp mill produced clean, absorbent paper that could be used for a throwaway diaper. US 3180335, applied for in 1961 from their Cincinnati headquarters, patented the original idea. The patent explained that existing diapers were either rectangular or contoured. The rectangular types had the disadvantage of the "bulk of material in the crotch area", which made them difficult to use, while the haphazard way in which the material was arranged often meant that the absorbent material was not utilized "to the fullest extent possible". The other type was in the shape of an hourglass. This was expensive to make, and some of the absorbent material had to be removed during manufacture. Both types also had the disadvantage of leaks. These problems, of course, were resolved in the patent. It consisted of a waterproof "hydrophobic" lining with an absorbent material which was made into a box pleat configuration, with side flaps folding over the pads. The (25) in Figure 1 is not what it seems: it represents a "multiplicity of plies of creped cellulose wadding". The new product was named Pampers® and was test marketed in Peoria, IL, in 1961. Kimberley-Clark's Huggies® appeared several years later, and there has been a technology battle between the two giants ever since, with constant improvements such as tapes, elastic leg guards and breathable side panels in the battle for market share. In the year 2001 Huggies® were ahead with 44% of the market, while Pampers® had 23%. Activists argue that improved reusable cloth diapers are better for the environment. The average child will use about 5,000 diapers and it takes some 800 lb of fluff pulp and 280 lb of plastic (including packaging) to supply one baby for a year. All of which normally goes to landfill.

Apart from inventions *for* children there have been many *by* them, some of which have been patented. Children have the advantage of looking at everything with fresh eyes, and not being scared to ask why something is done or "what if", although attendance at school may discourage such free-thinking. Jeanie Low of Houston invented a useful device while still in kindergarten. She had broken plastic stools when trying to reach up to the washbasin. To avoid that, and because stools could clutter up the floor, she suggested a "Folding step for cabinet doors", which was "particularly useful for small children in allowing them to reach a sink or counter top". It was secured by magnets and opened out when tugged. Her parents took her to a shop to purchase wood, and when she told the owner what she was planning to do he said that it would never work. She built it, aged five, and some years later, in 1990, applied for what became US 5094515 after encouragement by a patent attorney who like her belonged to the Houston Inventors Association. Jeanie has continued to invent. For example, she has designed and built a bathtub alarm that gives a warning when the tub starts to overflow or when a small child is in danger of drowning; a doormat with automatic brushes; and easy-grip doorknobs for people with arthritis.

The youngest child to apply successfully for a patent may be Robert Patch of Chevy Chase, MD, who in 1962 applied for US 3091888, a "Toy truck",

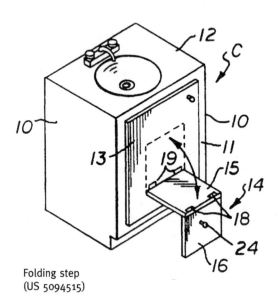

Folding step
(US 5094515)

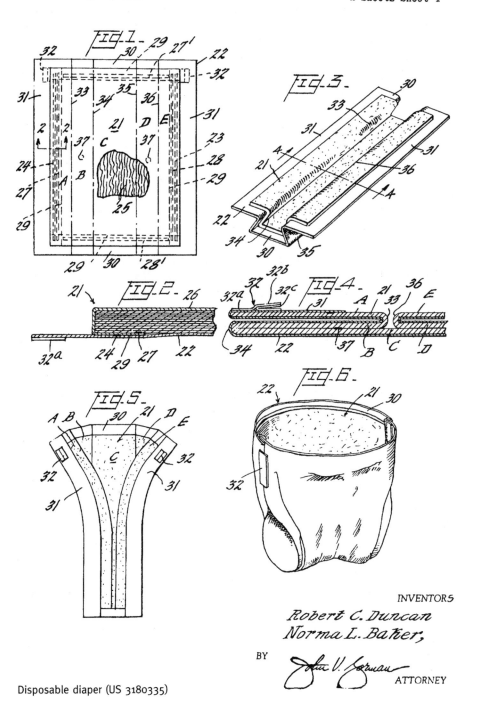

INVENTORS

Robert C. Duncan
Norma L. Baker;

BY

John V. Gorman

ATTORNEY

Disposable diaper (US 3180335)

aged six. The idea was to enable the truck to be changed from one model to another. Since ages are not given on patents it is difficult to verify claims. Another young inventor is Becky Schroeder of Toledo, OH, who from 1972 began to apply for patents when aged 10, 12 and 14 (depending on which website you look at). Her first patent was US 3832556 for a glowing backing sheet on which paper could be placed to enable writing in the dark. Other inventions by children include collapsible stands for school lockers; edible spoon-shaped crackers for cats; chalk dispensers; diapers with pockets for baby wipes; and solar-powered tepees. These patents are so well written that it can be assumed that some adult help was provided with the wording, but the invention itself is in theory at least by the child. The "Santa Claus detector" by Thomas Cane of San Rafael, CA, applied for in 1994 and which emerged as US 5523741, is not thought to be by a child, even if it is intriguingly child-like. It is for an electrical device which sets off a signal if anything is placed within a stocking.

Dolls have always held a special place in the hearts of many children (and quite a few grownups). One reason for their appeal must be the ability to manipulate the "lives" of miniature people, which must be instinctively attractive to those who must obey grownups. There are numerous individual dolls which have special places in individuals' memories (and perhaps households), with many of them scarcely known elsewhere. They show how often the appeal carries on beyond childhood to grownups, as is the case with so many "soft toys" today.

The first American patent for a doll is often cited on the internet as dating from 1840, but the first is actually from 1858. Ludwig Greiner of Philadelphia with his US 19770 explained how to make a less fragile doll's head. Being true to life has long been a goal, such as the 1991 Mattel filing which says, "One of the primary goals in providing a doll having a wetting feature is to add authenticity and realism to such dolls". The first wetting doll was by Marie Wittman of New York City with her US 1859485, filed in 1931. It was manufactured to some acclaim by the Effanbee Company as the Dy-Dee Baby. Wittman's later improvement, US 2007784, delicately mentioned that "an outlet provided at a suitable place in the body permits of the slow egress of said liquid into a cloth or diaper to thereby simulate a human function".

Another unusual doll patent dates from 1979, US 4249337 by Theresa Edson of Martinsville, IN. It is a "Breast feeding doll set". Then there is the crying doll by Mattel in 1974, US 3855729, with its "inflatable sack which pressurizes a tearing reservoir", and General Mills Fun Group of Minneapolis with US 3858352 in 1973, a "Doll with ingestion system". Food is taken in by closing a circuit which creates a pumping action within the doll. This is done by triggering the electric switch in the mouth by inserting a spoon or bottle. The food is later discharged at an "opening remote from the mouth". It may seem incredible that such patents exist but presumably they cater to a ready market. Probably the first really popular American doll was the Kewpie® doll, which was responsible for a world-wide craze for a couple of years. It was one of the first merchandizing opportunities to be fully exploited, as just about everything under the sun was sold with the name, including salt shakers and pianos. Women even plucked their eyebrows to imitate the cute dolls. And that was the odd thing about the dolls: they mostly sold to grownup women. Artist Rose O'Neill Wilson of Day, MO, applied for US D43680 for the doll in 1912.

A longer lasting success by another artist, which appealed to children as well as grownups, was the Raggedy Ann® doll. John Gruelle was born in 1880 in Arcola, IL, the son of an Indianapolis artist. He worked as a cartoonist for several newspapers and illustrated articles in magazines. The story goes that one day his daughter came into the studio with a battered rag doll which she had found in Granny's attic. Gruelle used his pen to give her a new face: round black eyes, a triangular nose and a wide grin. From a book of poetry behind his desk he took the titles of two of his favourite poems by his friend and fellow Hoosier James Whitcomb Riley—"The Raggedy Man" and "Little Orphant Annie"—and asked, "What if we call your new doll Raggedy Ann?" The story is so magical that even if it is not true it ought to be. There is a recollection in the family that a forgotten doll was indeed once retrieved from the attic, though the finder may have been Gruelle himself rather than Marcella. In May 1915 he applied from his New York home for a design patent, US D47789, for a doll, as well as a trademark for the name. It was a good touch to have the reminder of her name on the ribbon.

Tragically, Marcella died from an infected smallpox vaccination in November 1915, aged 13. Gruelle told Marcella stories about the doll in her last days, and in his grief he began to write them down. They were published as a book in 1918, *Raggedy Ann stories*. The human character in the stories is named Marcella, while Raggedy Ann has a candy heart to show her sweet nature. The book and the doll boosted each other's sales, and rumors that the dolls sometimes had real candy hearts sewn in did not hurt either. Raggedy Andy® came along in 1920 and was also registered as a trademark, although he does not seem to have been the subject of a design patent. After a campaign involving thousands of letters, Raggedy Ann® was admitted in March 2002 to the National Toy Hall of Fame, having previously been rejected four times. Gruelle's granddaughter, Joni Gruelle Wanamaker, spoke at the ceremony. "After September 11, people are looking for comfort and love and compassion", she said. "You can always cuddle a Raggedy Ann doll."

A more modern-looking doll is the famous Barbie®, over a billion of which have been sold across the world. She originated in a cartoon character called Lilli who appeared in the German newspaper *Bild-Zeitung*. A doll based on the character was launched in 1955, but the sexy teenager did not attract many sales in Germany. Rights to the copyright on her appearance were then purchased by Ruth Handler, the co-founder with her husband of Mattel in 1945. Her own daughter was called Barbara, while her brother was named Ken. Handler had noticed that her

Fig. 2   *Fig. 1.*

Raggedy Ann® doll (US D47789)

daughter and her friends liked playing with the kind of dolls sold to grownups. The idea of a teenage fashion doll (rather than a baby or toddler) was unique at the time. A designer was hired who created a wardrobe for the new doll.

The doll's launch in 1959 at the American Toy Fair in New York City met with scepticism from buyers working in the industry, but the public soon clamored for more (at $3 each). Sales reached $500 million in a decade. Some of Handler's family were not quite so keen. In 1989 daughter Barbara told the *Los Angeles Times* "I'm tired of being a Barbie doll", while her uncle Ken, a realtor, said of the doll,

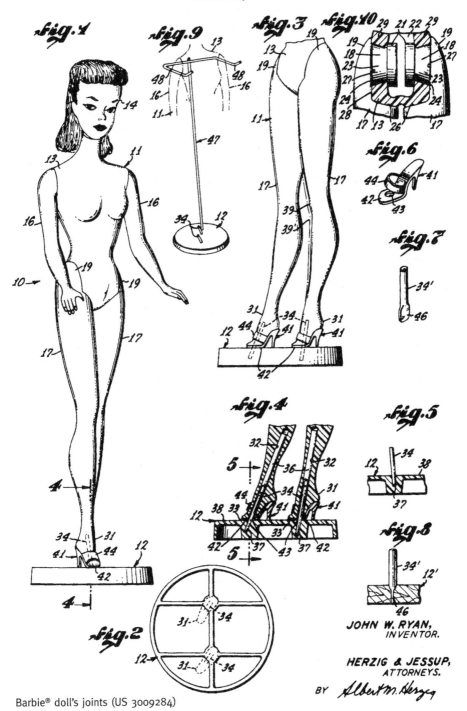

JOHN W. RYAN,
INVENTOR.

HERZIG & JESSUP,
ATTORNEYS.

BY Albert M. Herzig

Barbie® doll's joints (US 3009284)

"I don't really like her—she's a bimbo". There has been some controversy over Barbie®'s figure, since her original measurements were (it has been calculated) an impossible 36-18-38. This may make her a poor role model for many confused girls picking at their food. Sales are helped by her vast number of accessories, which include a wedding trousseau for her 40-year plus engagement to Ken. The drawing is from US 3009284, filed in 1959, which is for the method of articulating her joints, and which also displays her bland charms to good advantage. Barbie® could not actually bend her knees until 1965.

The inventor was John "Jack" Ryan of Bel Air, CA. Earlier he had worked as an engineer designing Sparrow and Hawk missiles. Just as improbably, he was the sixth of Zsa Zsa Gabor's eight husbands. Mattel are said to have hired him for his knowledge of modern materials. Ryan was also responsible for the technology in the original Chatty Cathy® doll in 1960, who could speak 11 phrases at random using a 3-inch phonograph record concealed inside her, with US 3017187.

Just as Barbie was a big hit with girls so was GI Joe® with boys: the first successful doll (although never called that) for boys. GI Joe® was inspired in part by a television show called *The Lieutenant*. A licensing agent named Stan Weston brought the concept to Don Levine with the idea of creating a toy soldier for boys. Levine and his designers liked the idea. US 2377602 was filed in 1964 by Hassenfeld Brothers (now Hasbro) of Pawtucket, RI.

As with the Barbie® patent the invention is in the way the joints work rather than the general appearance. He has the right tough jaw, but is normally seen clothed. GI Joe® was sold in four versions with over 70 different accessories as equipment: he was a soldier, sailor, Marine and pilot. There were huge sales for several years and variants were introduced at intervals. One change was that the original hard hands, which could hardly grip anything let alone a gun, were altered to softer, more flexible hands. Then sales dipped and a change was made with the Adventure Team, where the emphasis was on saving animals and the environment. Sales were better than ever until production stopped in 1976. In 1982 a new line was introduced, Real American Heroes, with smaller sizes of under 4 inches to match the familiar Star Wars characters. Many more variations have been tried since, some trying to capture passing fads such as the Ninja craze.

Increasingly, perhaps because of our electronic age, children expect dolls that can interact with their owner. One example is the Furby® toy that apparently talks to you, and which gradually incorporates English into its Furbish dialect. The patent by Tiger Electronics of Pawtucket, RI, US 6149490, is 59 pages long including 42 pages of drawings. This includes discussion of the need for a sturdy, compact toy that tolerates being dropped. Two designs were also applied for, showing Furby® as he is normally seen and (clearly taking no chances of illegal copying) with the fur removed. The ears, eyes and mouth are all moveable, powered by a small motor which moves a 1-inch-long control shaft from which all the mechanisms operate on the cam principle (wheels turn so that other parts will move). A

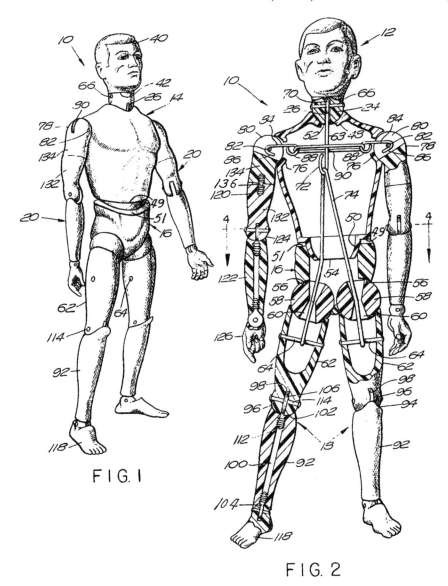

FIG. 1

FIG. 2

GI Joe® doll's joints (US 3277602)

problem with interactive toys for children, apparently, is that the child's speech is often very difficult for the computer in the toy to comprehend and hence to respond to appropriately. That is why so often there is the "cut and paste" idea of playing back what has just been said. In the Furby® soft toy a lot of thought by the designers has gone into making conversations evolve.

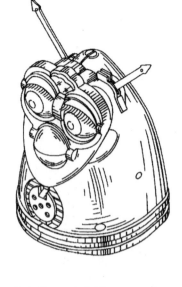

Furby® clothed (US D423611)   Furby® revealed (US D419209)

Electronics has indeed moved firmly into the toy industry with Barbie® herself having joined the computer age. The Barbie Fashion Designer game was launched in 1995 and quickly sold a million sets. It was the first successful electronic game to be aimed at girls. The game starts with the words, "Hi, I'm Barbie. Let's make some fun clothes for me to wear". Outfits that can be worn by the doll are designed and printed out on fabric covered paper that fits in an inkjet printer. The game— if you can call it that—annoys feminists and hard-core gamers but at least is creative, in the same way that the Sim City® series of computer games is creative. The drawing from US 6206750 dates from 1998 and is of a later version. No fewer than 12 inventors were credited.

Genetic engineering, including the cloning of embryos, is highly controversial. The cloning of tissue from a dead child so that its identical twin may later be born is one example. US 6211429, which was published in April 2001 by the University of Missouri, falls into this field. The patent covers a way of turning unfertilized eggs into embryos, and the production of cloned mammals using that technique. Unlike other patents on animal cloning it does not specifically exclude humans from the definition of mammals, and specifically mentions the use of human eggs. Those opposed to cloning and to patenting living things say the patent is a further sign that human life is being turned into a commodity. "It is horrendous that we would define all of human life as biological machines that can be cloned, manu-factured and patented", said Andrew Kimbrell, executive director of the

International Center for Technology Assessment (CTA), a Washington, DC, group that opposes the patenting of living things and which wants to ban human cloning. Senator Sam Brownback, the Kansas Republican who is a leading opponent of human cloning, said patents on human beings and human embryos were "akin to slavery".

The Patent and Trademark Office has long forbidden the patenting of humans, body parts or human embryos. Many experts say this is because such patents would violate the 13th Amendment ban on slavery. Brigid Quinn, a spokeswoman for the Patent Office, said the agency was not using the 13th Amendment argument anymore but was not granting patents on humans because it had not received any guidance from Congress or the courts saying it should do so. "We do not patent human beings." Russell Prather, one of the university inventors, working with BioTherapeutics, has used his method to clone a miniature pig that was missing a specific gene. The advance was reported in the journal *Science* and may help make animal-to-human transplants a reality. But CTA said the patent allowed a "method for producing a cloned mammal". Since humans are mammals, the researcher could use the method to clone humans, said Joseph Mendelson, legal director of the CTA. Because of the debate, "University officials plan to carefully review the

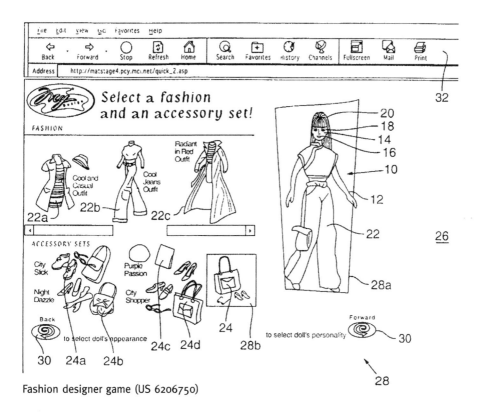

Fashion designer game (US 6206750)

language of the patent", according to a statement issued by the university. In the meantime, CTA is urging Congress to rewrite patent laws to prevent three similar applications from being granted. These are from a company and two universities. Money, of course, is powerful, and who knows what the future will hold.

"The play's the thing"

THERE are tens of thousands of games and toys among the American patents, most of which never made it to the marketplace. One reason for the vast numbers is that toy manufacturers reckon on a new generation every four years who must be provided with fresh variations. Most of the patented toys, especially by private inventors, were never manufactured, and few even of those were ever a success. Many only survive in a few attics, or in the lives and websites of devotees. Many concepts come from abroad, with some well known games deriving from Britain or France. Patented board games include the British-origin games Clue® (only patented in Britain, where it is called Cluedo®), Careers® and Sorry®, while Stratego® and Risk® were patented only in their native France and in Britain. Others were never patented at all such as Trivial pursuit® (which is Canadian).

Most early games were religious, moralistic or at least educational, as games played simply for fun were considered rather immoral. It is no accident, for example, that so many early jigsaw puzzles involved geography. Some games are still advertised as educational, while others downplay such an angle, perhaps because few children ever plead to be given an educational game—although they might enjoy it if it is simply another fun game. At the most basic level traditional games have the virtue of encouraging "interaction" with others, rather than with computers.

It is possible to estimate the approximate number of patents on toys and games to 1873, as a keyword index exists for those years. Of over 155,000 patents issued some 400 were listed as beginning with "toy" or "game". A typical example (if that is possible among so many different inventions) was by James Swain of Philadelphia, who filed in 1854 for a "Magnetic toy called the magical Cupid". Published as US 10423, it was an early fortune-telling game, "intended for amusement and instruction". Illustrated is an example of one of the supplied question "blocks" which was inserted into the box and pressed against a pin (Z) which in turn affected the magnet below Cupid, who swung in response to one of the possible answers. The answers were in fact predetermined. The trick was having tiny holes in each block (to press against the pin) drilled to varying depths, so that the pin would generate set responses to each question. The spring was used to eject the blocks. Sadly, although Swaine suggested two other questions in his patent, he did not supply a list of the prescribed answers to suit his drawing (as the questions were only examples). Although the toy is ingenious, purchasers of it (if it were ever manufactured) would soon have tired of the unvarying response to each question.

The first American board game is sometimes credited as being the "Traveller's Tour", marketed by the book publisher Lockwood's in 1822. This honor is more

Fig: 1.

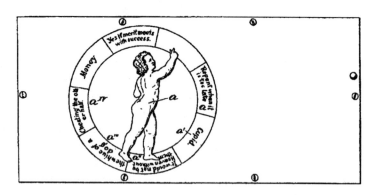

Yes if merit meets with success.

Money

Forget when it is too late.

Cheating the old crab.

Cupid

It would not be Heaven without a dog.

The value of a Derm.

a a' a'' a''' a^{iv}

Fig: 3.

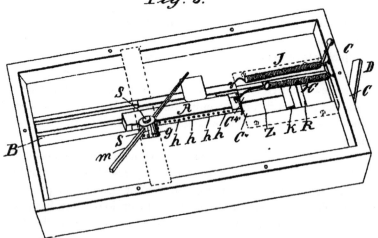

Fig: 4.

What is one half the World doing?

Magical Cupid toy (US 10423)

frequently attributed to "The Mansion of Happiness, An Instructive Moral and Entertaining Amusement", which was published in 1843 by W. & S.B. Ives of Salem, MA. The inventor was Anne Abbott, a clergyman's daughter from nearby Beverly. A highly moral game, it involved the players moving along a spiral track through both virtues and vices such as honesty and idleness while being punished at, for example, the whipping post before reaching Happiness in the center of the board. It apparently sold well.

Many early games involved cards. The educational aspect is so blatant, and they are so boring, that it is astonishing that (as presumably often happened) these games ever sold. One example was by Reuben Sitterley of Carthage, MO, who filed in 1876 for a geographical card game, US 183335. The object was "to provide a means whereby amusement is combined with instruction in the rudiments of the science of geography and history". The California card for example gives the date of settlement as 1769 (apparently the Indians did not count) together with some dull facts. Before play the map sheets were inserted in the rack according to the numbers on their back, so that, for example, Nevada was number seven. The players were then dealt the cards giving the facts on each state. A nominated player would call, say, the size of the state as the factor for the ensuing game, and might play his card for Indiana. If a player held Ohio this could be played to defeat Indiana, as it is larger in size. Each time a card was played the corresponding map sheet was removed from the rack and added to the jigsaw provided that it bordered the previous state set down. The aim was to have the largest number of tricks from captured cards when play ceased. There were other rules, but this was the essence of the game.

Meanwhile another card game had been invented by a 16-year old. George Parker of Salem, MA, formed a club to play games with his friends. He devised a card game called Banking where the players borrowed money at ruinous interest rates and hoped to make a profit after paying the interest back. Efforts to make money were not considered to be immoral. 160 cards determined the investors' fates. His friends liked it and suggested that he sell it to the public. Parker tried two Boston publishers, but they both turned him down. He spent $40 of his life savings in having 500 sets printed and spent his last $10 in expenses in hawking his game around during 1883. At the end of the year he had only a few dozen sets left, and a profit of $100. He started a games business and in 1888 persuaded his brother Charles, an oilman, to join him to form Parker Brothers. A third brother, Edward, joined in 1898. There have been 1,800 games since, with most of the early ones invented by George, who also arranged for advertising, unheard of at the time.

Slot cars are the little cars that race around a track. The idea is credited to Albert Cullen of Willow Grove, PA, who filed in 1936 for US 2112072. Power from a battery (C) went from the connector (D) via a "controlling switch" (29) to a conductor running down the groove (18) in the center of the track, and was picked up

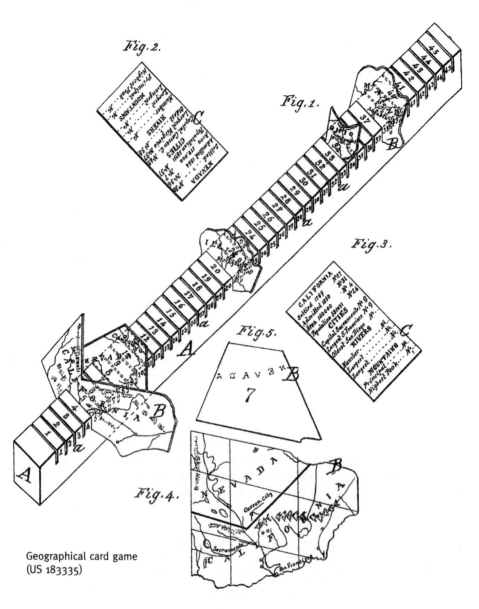

Geographical card game
(US 183335)

by the "roller" (43) shown at the front of the car. This powered the little electric motor on each car and also controlled the car's movements. The track was preferably to be of metal and could be colored to look like asphalt. The idea of parallel tracks so that cars could race against each other was a later development, although Cullen did suggest that the track could form intersections (which are more fun if there is more than one car on the track, with the risk of collisions). It is said that

Cullen did not make any money from his invention as he refused the small sum offered him for it.

There have been so many board games patented in the past which no longer have any protection that games manufacturers would find it useful going through the old patents looking for ideas (and perhaps do). Here are a couple of examples of possible or known derivations from old patents.

US 836537 was filed for in 1906 by William Roche of New Bethlehem, PA. It has a delightful period feel and involved playing cards to advance between 1 to 10 miles, depending on the cards in your hand, or inflicting on other players a flat tire (given in the drawing as "puncture", the British word), being out of gasoline or being involved in a collision. This is a concept found in many games: do you play to advance yourself, or do you use your turn to block others?

Other cards enabled these problems to be overcome by supplying spare tires and so on. Both cars and roads were such at the time that these were genuine problems that every wise driver prepared for by having a tool kit and supplies in the auto-mobile. Collisions may have been rare since the speed limit card that can be played

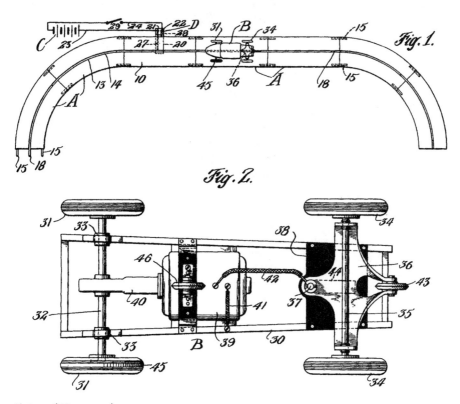

Slot car (US 2112072)

No. 836,537.

PATENTED NOV. 20, 1906.

W. J. ROCHE.
GAME.
APPLICATION FILED AUG. 14, 1906.

2 SHEETS—SHEET 1

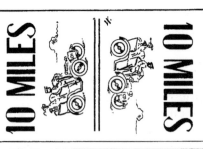

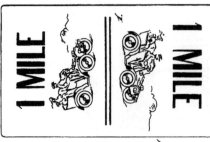

Motorist card game (US 836537)

was set at five miles an hour. In this case, too, a very similar game was launched by Parker Brothers in 1962, Mille Bornes®. The name is French for "thousand milestones", and the wording on the cards is also in French, which perhaps conveys an attractive foreign flavor to the game as you struggle to cover 1,000 miles. As cars are now faster, some cards in the modern game allow you to cover 200 miles in one turn. Internet sources credit Edmond Dujardin of France as being the originator in 1960. The similarities between the Roche game and the modern one may just be coincidence.

Another game was first patented by Lizzie Magie of Brentwood, MD, in 1903 with her "Landlord's game" being published as US 748626. The object was to make as much money as possible while playing a fixed number of circuits of the board. Each player started with $500 and 12 of the 22 "lot tickets" for (unnamed) properties were shared out between them. Each lot was marked by its rent and price for its sale. Four unnamed railroads were there in their well-known places, each rented for $5. When a player passed the "Mother Earth" corner square $100 was allocated as "each time a player goes around the board he is supposed to have performed so much labor upon mother earth". The other corner squares were Go to Jail, Public Park/Poor House and Jail. She clearly did not like the British: the Go to Jail square is also labelled "Owner Lord Blueblood London England No trespassing".

Magie was a supporter of Henry George, a Philadelphia-born radical who became a Californian by adoption. His book *Progress and poverty*, published in 1879, proposed a single federal tax based on land ownership (not the actual buildings, and with no taxes on wages or interest). Such a tax would, he hoped, discourage speculation and encourage equal opportunity. The game was meant to be a teaching aid for George's ideas. The game spread by supporters painting the board on tablecloth, often with their local city's street names. In 1923 she applied for a revised version, US 1509312, which in some ways was closer to the ideas in Monopoly® and in others was further away. The patent was in her married name of Elizabeth Magie Phillips.

The goal was to be the first to win $3000. The addition of place names added a certain flavor to the game, as countless variations over the decades have shown. Some land was in use and could be bought or sold while other lots were lying idle. Chips were introduced for the first time which could be bought to improve the value of properties. "Lord Blueblood's Estate" was there again, this time as a property lot, to symbolise "foreign ownership of American soil" while "La Swelle Hotel" showed the distinction between classes, "with moneyed guests only being accepted". Owning all three utilities or all three railroads enabled higher rents but this did not apply to the other lots.

Charles Darrow's US 2026082 for Monopoly®, which he filed in 1935, did have significant differences such as complete sets of properties meaning higher rents (although Phillip's second version did have different colors for different types of

Forerunner of Monopoly® board game (US 1509312)

property). Parker Brothers bought her out in order to ensure that they could manufacture the Darrow version, whose less political version would have sounded more saleable. A reporter for the *Washington Star* wrote about Magie's game in a story published on January 28, 1936. The reporter asked Mrs. Phillips how she felt about getting only $500 for her patent from Parker Brothers and no royalties. She replied that it was all right with her if she never made a dime so long as the Henry George single tax idea was spread to the people of the country. It is unlikely that many players of Monopoly® have, in fact, ever heard of this tax reformer or been inspired by his ideas. Early sets did credit both her patent and Darrow's.

There is in contrast no doubt about the origins of the famous Erector® set. Alfred Gilbert applied in 1913 on behalf of the Mysto Manufacturing Company, both of New Haven, CT, with his US 1066809. His "toy construction blocks" were meant to simulate steel construction, something that fascinated him at building sites. Bolts and clips to secure the "steel beams" were also provided for.

Gilbert held some 150 patents, mostly for his A.C. Gilbert Company. He was a doctor and had also won the gold medal for the pole vault in the 1908 Olympics. He supported efforts to encourage the buying of toys at times other than at Christmas and became the first president of the Toy Manufacturers of America when it was formed in 1916. In this role he soon had to lobby on behalf of the industry. When the USA entered World War I in April 1917, the Council for National Defense proposed prohibiting the selling of toys in an attempt to conserve resources. Gilbert and a group of manufacturers set out for Washington, DC, with many of their products to meet the Council.

The *Boston Post* reported that they made an eloquent defense of the role of toys in modelling the patriotic character of America's youth. Gilbert then distributed the toys which led to what must have been an interesting scene. "The secretaries were boys again. Secretary of the Navy Daniels was as pleased with an Ives submarine as he would have been with a new destroyer . . . he kept fast hold of it. . . ." Another member of the cabinet said, "Toys appeal to the heart of every one of us, no matter how old we are". And so the Council reversed its decision.

Another well-known toy was by John Lloyd Wright of Chicago, one of the sons of the famous Frank Lloyd Wright, and also an architect. He applied for US 1351086 in 1920 for the Lincoln logs® toy cabins with logs that can be assembled

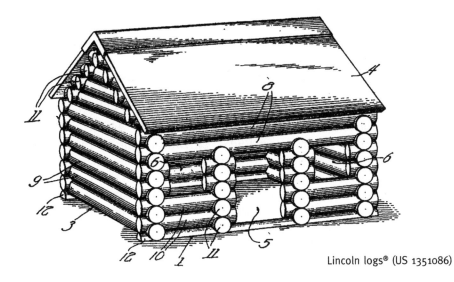

Lincoln logs® (US 1351086)

to interlock with each other. It is said to have been inspired by his father's design for the Imperial Hotel in Tokyo, which was built in 1916–22. This was meant to be earthquake-proof, and indeed survived the great Tokyo earthquake of 1923. Lincoln logs® is a very appropriate invention for an architect.

The foreign (and British, as it happens) origins of two well known games are by contrast well established. William Henry Storey of Southend, England applied for US 1903661 in 1930. It is an example of a game which is almost forgotten in its country of origin but which is much loved in the USA: Sorry !®. Storey himself was enterprising enough to register the trademark in the USA in 1930 and he licensed the game to Parker Brothers in 1934. The original box calls it "The fashionable English game". The modern board is only slightly modified from the original.

There are four "start" tracks marked as (3), which join the main track, and also four longer tracks (6) which end in "homes" (7). These tracks are in four different colors. Each player owns a "start", "home" and four "men" or counters, all of the same color. The object is to move all four counters from the "start" to the "home" as influenced by the 44 cards in the pack. Instead of dice, the cards indicate the number of moves along the tracks. Picking up a "Sorry" card meant that the player could move a counter from the "start" to the square occupied by an opponent, whose counter had to return to his or her own "start". Storey stated in the patent that the game was proved by experience to be "a prolific source of amusement" requiring much judgment (as it is best to delay moving all your counters from the "start" square).

Clue® (or Cluedo® in its native Britain) is also a much loved board game. Under British law only the board and equipment can be patented, and not the rules (American law is more relaxed) and this game certainly has plenty of equipment. Anthony Pratt was a law clerk from Birmingham, England who had already patented a card game and this was his only other patent, GB 586817. He only patented it in Britain but as he applied for it in 1944, during difficult times for the British, that is hardly surprising. He loved murder mysteries and thought up the idea (supposedly while while keeping watch for fires from bombing) while his wife designed the board. They tried the game out with their friends, who liked it. Pratt approached the British games manufacturers Waddington's. Their executives sat down with him, also liked it, changed his suggested name of "Murder" to Cluedo® (to benefit from their trademark for Ludo®, the British name for Parchesi) and agreed to manufacture it. Parker Brothers licensed the American rights and applied for the trademark Clue® in 1948.

The following explanation is not just for the benefit of those who have never played the game: it is based on the patent's explanations of how to play it, some of which have been modified over the years. The players are detectives who must solve a murder: who was the murderer, with what weapon, and in what room (the last two would surely be obvious in most cases, but never mind). There were ten

Sorry !® board game (US 1903661)

characters: Dr Black, Mr Brown, Mr Gold, Rev. Green, Miss Grey, Prof. Plum, Miss Scarlet, Nurse White, Mrs Silver and Colonel Yellow. The nine weapons are illustrated but only four are named (". . . different weapons or poisons such as axe, poker, rope . . .") with bomb, hypodermic needle, revolver, knife and a candlestick also being available. The house was much grander than modern versions, with nine rooms that could be used for the murder (the gun room and cellar were apparently off limits). The three packs of cards (people, weapons, rooms) were shuffled in separate packs. The top card from each of the three packs were hidden under the board. Deduction must be used to identify the correct combination of murderer, weapon and room. The cards are placed face down on the various rooms.

From their starting places at (15) on the board the players, who use different sized pieces if male or female, move into the rooms through the doorways (14) to

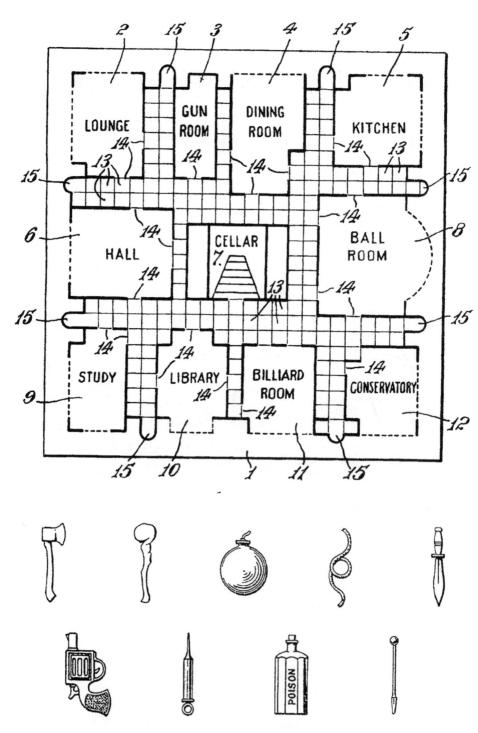

Clue® board game (GB 586817)

pick up cards. When all cards have been picked up the players move their pieces into rooms to make accusations: the murder occurred here with a certain weapon by a certain person. Each player must privately show the accuser any named cards. Gradually a player eliminates suspects until the winner is able to name cards not held by anyone.

Waddington's (now owned by Hasbro) were so pleased when they reached 150 million sets sold that they decided to award a trophy to Mr Pratt. Unfortunately the patent had long expired, and nobody knew where he was. A hotline was set up for information and eventually a cemetery superintendent phoned up to say that he had died two years before, in 1994. His tombstone says "Inventor of Cluedo".

One of the most familiar names in American toys is surely Fisher-Price (now part of Mattel). The company specialized at first in "pull-along" toys where the pulling or pushing actuated something, such as a 1950s Looky Fire Truck, where the two firemen at the back turn as it is towed along (sadly not patented). These toys were for a variety of creatures and objects and many involved Disney characters, such as US 2424607, from 1945, which was for Donald Duck towing his nephews along. The very first patent by the company, US 2069181, is from 1936 and was by Herman Fisher himself of East Aurora, NY. It is for a mouse beating a drum and the patent explains how the turning of the wheels causes the mouse's arms to beat the drum. The drawings make the mechanism clear with Figure 3 showing the toy from below.

Herman Fisher founded Fisher-Price Toys and served as president and chairman from 1930 until 1969. The company specializes in preschool toys. Fisher believed that all of his toys should have intrinsic play value, ingenuity, strong construction, value for money, and action features.

WHAM-O is a company that has been credited with marketing two of the greatest successes in toys. In both cases plastics were vital in making the products a success. In 1957 a visiting Australian mentioned to Californians Richard Knerr and Arthur "Spud" Melin, the founders of WHAM-O, that in his country children twirled bamboo hoops around their waists in gym class. This sounded interesting, and they made a hollow plastic hoop which sold 25 million at $1.98 each in two months. The craze died down by the end of 1958 but not before much money had changed hands, and a lot of children had briefly at least become very happy. The plastic used was Marlex®, a light and durable plastic made by Phillips Petroleum.

WHAM-O's other big success was more enduring. The origin of the Frisbee® flying disc seems to date back to at least 1920 when a Yale student recalled the throwing of pie tins from the Frisbie Baking Company of Bridgeport, CT. It was not until 1948 that Walter "Fred" Morrison and Warren Francioni, intrigued by the flying saucer phenomenon, made the first plastic flying disc. This had better qualities than metal, although the plastic he initially used was so brittle that it easily snapped. Morrison tried to market the product with trade names such as the Arcuate Vane, the Rotary Fingernail Clipper, the Pipco Crash and finally the Pluto

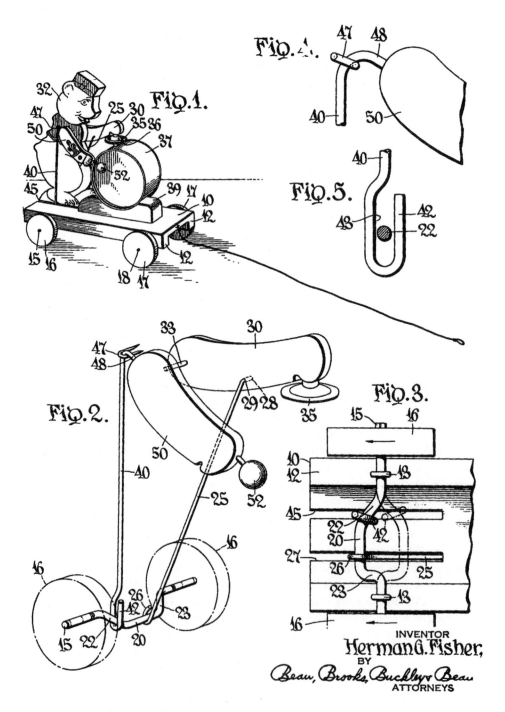

FIG.4.

FIG.1.

FIG.5.

FIG.2.

FIG.3.

INVENTOR
Herman G. Fisher,
BY
Beau, Brooks, Buckley & Beau
ATTORNEYS

Mechanical toy (US 2069181)

Platter. This involved a 1951 redesign with the curved "Morrison slope" as the rim which all such discs have. He demonstrated it on beaches and on parking lots, at one of which he was noticed in 1955 by Knerr and Melin, who invited Morrison to join WHAM-O. They began producing the Pluto Platter in January 1957 and Morrison filed for US D183626, a "flying toy", in July 1957. It included little circles for flying saucer portholes (he was still very interested in the subject). The success of the hula hoop diverted the company's attention, and it was not until 1958 that they relaunched the product as the Frisbee® flying disc. It is thought that the name came from noticing the warning call "Frisbie !" on a visit to east coast campuses.

It is easy to see its flight but much more difficult to explain how the "rotating disc-wing" flies as it does. Four things are needed for it to work. One, an atmosphere, obviously. Two, force in the initial throw. Three, rotation. Four, it has to be flicked with the hollow side downwards. Skill from practice helps, of course, especially for direction. Filing a design was odd as they are for looks and not for how something works. A patent *was* filed, but not until 1965. Edward Headrick for the company filed for US 3359678 which was for the thin lines on the top of the disk plus the thicker edges.

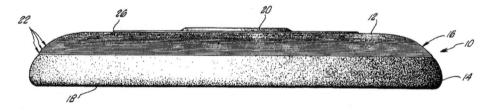

Frisbee® flying toy (US 3359678)

The aerodynamics were perfectly centered so that the disc could maintain its orientation for a long time as it has angular momentum, which dramatically changes the way it responds to what is called aerodynamic torque. The patent improved performance by thicker edges on the disk which maximized its angular momentum when it spun, and by the tiny ridges on the disc's top surface which introduced microscopic turbulence into the layer of air above. This turbulence helps to keep the upper airstream attached to the disc, so that it can travel farther. A different design, and especially throwing it upside down, would give a Frisbee® disk the flying qualities of a brick.

Some products are thought to be secret, and one is the formula for Play-Doh® modeling clay. It was in fact patented with US 3167440, filed in 1958 by Noah and Joseph McVicker for Rainbow Crafts, all of Cincinnati, OH. The trademark had been used from 1955. The clay was originally white, and colors were added later to make it look more interesting. It sold in huge quantities. The patent describes the recipe which consisted of flour (preferably wheat), water, a hydrocarbon distillate such as kerosene and a "soluble saline extender". Other ingredients such as perfume and coloring could be added.

Many must be grateful for the invention of Etch A Sketch®, which went on sale in 1960 and sold 50 million in the following decades. It was sold by the Ohio Art Company, also of Cincinnati, which was better known for metal or lithographed products. The toy was designed by Arthur Grandjean for Paul Chaze, both of France. It involves a rectangular television screen-type image with two knobs that move lines up and down. "Pulverulent" (aluminum) powder adheres to the inside of the case and twisting the knobs removes it as lines are drawn within the casing. When turned upside down and shaken, the lines disappear as the material returns. Bravely, the patent describes itself as an "educational game". The only change has been replacing the old brass knobs with larger plastic knobs. Who should be grateful for this invention? The game is popular with mothers because it has no accessories, batteries or noise.

Also popular is the Mousetrap® game. Marvin Glass and Gordon Barlow of Chicago, IL, applied in 1962 for US 3298692 which has the title "Game with action producing components". The apparatus is assembled piece-meal as a result of the players' progress round the board. The idea is that once all the components have been assembled, the Rube Goldberg machinery in the middle is set in action and, if correctly assembled, traps the mouse. Glass stated that the game would "appeal to younger players" and hoped for educational effects from their having to put together the parts. Glass had, when still in his twenties, formed in 1941 Marvin Glass & Associates, the first (and greatest) "toy studio" to create new toys. Some 75 people worked there. The toys were sold on as developed products to manufacturers. For many years their logo was printed on the boxes of their games even though they were not the manufacturers.

The company was regarded as the Walt Disney of toy making, but with a distinctly maverick touch. Nothing was too wacky to include in their range. The company had over 700 patents, some 200 by Glass himself. The better-known products included Mr. Machine, Kissy Doll and Rock-Em-Sock-Em Robots™. Lesser known products included Yakity-Yak talking teeth and Super Specs. This last product consisted of gigantic sunglasses which were modelled on the box by a cigar-smoking Glass. The story goes that he had been at a party and while listening to a bore droning on imagined the man's spectacles getting bigger and bigger.

Glass was obsessed with the thought that his company might be spied on from the Moody Bible Institute across the street from their offices on North LaSalle Street, Chicago. Therefore his office had double walls and looked out onto an inner courtyard, and all toys and games being worked on had to be locked in vaults each night. This practice was discontinued after Glass' death in 1974, and the company eventually closed down as staff formed their own studios.

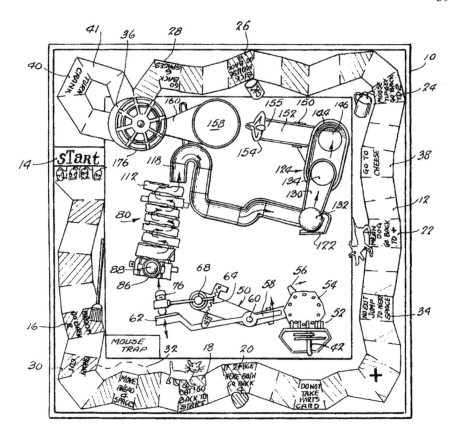

Mousetrap® game (US 3298692)

The idea of toys that can readily change their shape or appearance by, for example, being turned upside down has been known for a long time, but transformer toys took the concept much further. It was Japanese work which led the way from the mid 1980s, with Hasbro being the leading American company to be involved. Early transformers could change into realistic looking trucks, cars, planes and household objects such as cassettes or a Walther P-38. Futuristic-looking vehicles and mechanical animals came later together with transformers whose heads and weapons were themselves transformers.

There were teams of transformers that merged to form giants (known as *gesalts*) and there were also virtual toy cities. The largest of these is thought to be Fortress Maximus, which is shown in US D305786, which was filed in 1987 by the Takara Company of Japan. From its ten pages of drawings are shown views as a robot and

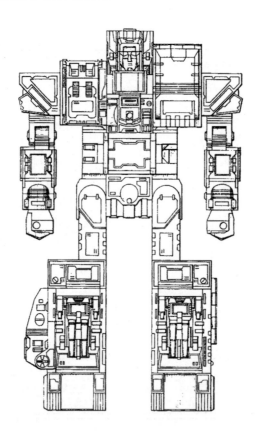

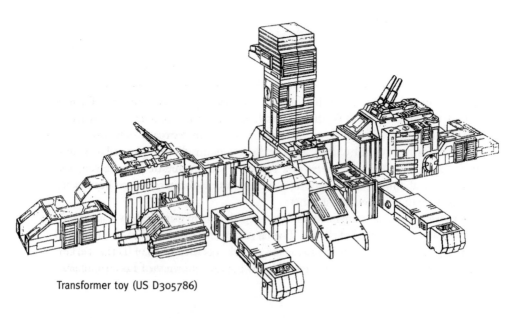

Transformer toy (US D305786)

as a fortress. It was about 30 inches high and its head transformed into a robot which itself had a head that would transform into a robot.

Jurassic Park was released in 1993 and this must have helped the popularity of Beast Wars® which was launched by Hasbro in the mid 1990s. As a departure from earlier transformers, it could be changed into realistic-looking animals with fur and scales. The idea is that the evil Predacons are using vehicular transformers called Vehicons to take over the earth, while the good Maximals struggle to survive. Some older fans initially rejected them, but the line proved to be popular. Many cartoons and comics have taken on similar themes with Pokemon® (pocket monsters) just being one such.

Some new games come about because players are dissatisfied with something in an existing game. In 1987 three students at Colgate University in New York state, Dave Yearick, Ed Muccini, and Tim Walsh, heard that Trivial Pursuit®, had been invented by (Canadian) students at the same university. They confirmed this and in talking about its great success wondered about improving the game's major defect: either you knew the answer to the general knowledge questions or you did not.

Later, while Muccini was working in a pet shop, he thought of the concept for the game and talked to the others about it. It would be a game where the questions were clues to the answer. The idea lay dormant until two years later when, on a trip to Florida, they worked out in detail the concept. They formed a company, raised money from family and friends, and hired a company to make the first 2,500 sets. They tried licensing the game to Milton Bradley or to Parker Brothers but both rejected it, as did other manufacturers. A common response was that the American public did not want a game that would make them think. So they teamed up with the game's printers, Patch Products. They could not afford an advertising campaign, and instead asked disk jockeys to play the game with their listeners, and to give the sets as prizes. After more than two years of selling the game out of car trunks they managed to get it into major retailers, and by Christmas 1996 a million sets had been sold. Some of the same companies that had rejected them tried (unsuccessfully) to woo them back.

The game is called Tribond® because three ideas are linked. Hence in reply to what Florida, a locksmith and a piano have in common the answer is "keys".

The triangular game board continued the theme of threes by each player or team choosing one each of three scholars, saints and soldiers (there are of course three inventors). When all three characters have been moved to the center by correctly answering questions in various categories, the final clues have to be answered to win.

Similarly Scrabble® (patented as US 2752158) was modified by Upwords®, where you can change words by building upwards to a maximum of five tiles, and points are counted for each tile in the altered words. Elliot Rudell of Torrance, CA, filed for his US 4776597 in 1982 and Milton Bradley (now part of Hasbro) market

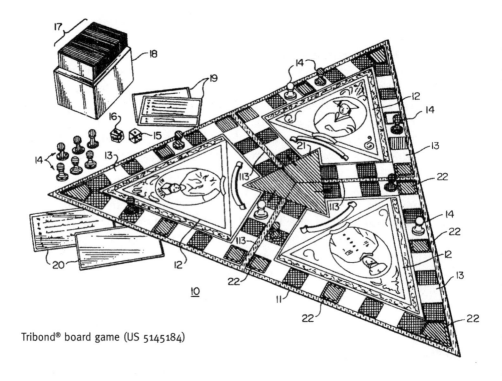

Tribond® board game (US 5145184)

the game—as they do Scrabble® itself. The new game means that many more possibilities are open, and there is no longer the need to watch out for the triple or double letter or word score. Adding an S just to make a plural is forbidden, though.

Such word games show that many games are played by grownups as well as by the young. Toys, too, can be for grownups, with "executive" toys a growing market. Quite how a busy executive ever finds the time to play with them is not clear. Many involve scientific concepts such as Newton's balls (the balls on strings that knock against each other) and the Dippy Bird (or Drinking Bird, or Dunking Bird), the nodding bird that relies on evaporation to power itself as a simple heat engine. This was filed in 1945 by Miles Sullivan of Washington, DC, as US 2402463. Interestingly, US 6343788, a "multistable mechanical switching device", mentions that it can be used as an "executive pacifier" among many other purposes. The most common type is probably those involving getting something to balance, or to manoeuvre little balls into the right pocket or around a circuit.

The sporting life

···

MANY people live to watch and, perhaps, even to play a sport. You have to support your side while consuming large quantities of pretzels, hot dogs, beer and anything else you fancy. Americans are devoted to their sports in a way with which Europeans (other than hooligans) are only just catching up.

Which is the most popular sport in America? It depends how you measure it. The contest seems to be among baseball, basketball and (American) football and baseball would seem to win on attendance, even though college basketball and football boost their figures. Basketball, though, wins in terms of the numbers who themselves played at least once in the year (29 million). On turning to watching on television the picture changes. In a 2002 Gallup poll asking which was their favorite sport to watch, 28% said (American) Football, 16% basketball, and only 12% baseball. Auto racing and golf came next. Only 2% said soccer. Nevertheless the drawing on page 44 depicts baseball as "Our national ball game", which it probably was at the time. Edward McGill of Philadelphia, PA, filed for his US 367991 in 1887. There are thousands of board games which reproduce the action of many different sports. Most of the older games involved chance by either using a spinner or (as in this case) a pair of dice. Skill was more difficult to introduce, although some board games tried, typically by flicking a ball towards a board marked with the results of hits (US 1509270 from 1921 is an attractive baseball game of that sort). The drawing shown includes odd phrasing: one, two and three base hits are mentioned instead of single, double and triple. A home run is 1 of the 21 possible outcomes of any "hit" even though they were rare at the time. Billy O'Brien hit the most home runs in that same year of 1887, for the Washington Senators, with a mere 19.

The catcher in the drawing on page 45 is shown without a mask. Catchers, of course, are more vulnerable than any other baseball player as they are always in the firing line. Frederick Thayer, the captain of the Harvard team, had in fact patented the first mask in 1878 with his US 200358. He made it by modifying a fencer's mask for the team's Alexander Tyng, who only had two errors when first using it in a game on April 12, 1877, which was excellent for the time. At first others called it a rat-trap, but its usefulness soon caught on, and Spalding sold it in their catalogue from 1878. Previously catchers had used rubber mouthguards such as those worn by boxers, an idea introduced by Harry Wright, founder of the Cincinnati Red Stockings.

Similarly fielding was done barehanded (as is still the case in cricket). As they tend to deal with the most powerful hits or throws, it was the first basemen and catchers who began the move to wearing gloves. Fingerless gloves that just covered

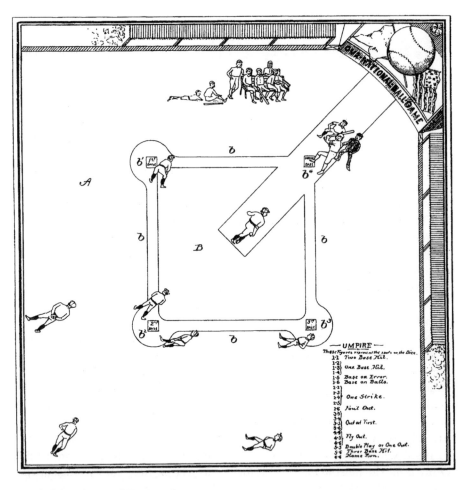

Baseball board game (US 367991)

the palm came first. Charles Waite of St Louis is thought to have been the first to use one, but when he took the field for the first time in 1875 with thin flesh–colored gloves, the fans (and even his own side) called him a "sissy". A little later Albert Spalding switched from pitching (for which he is a Hall of Famer) to playing first base. He decided that he would wear a black kid glove. Though it did contain some padding, his glove was closer to modern golf gloves than to a mitt. Spalding was so well known and respected that the idea was adopted by others. As the leading figure in A.G. Spalding & Brothers he also turned the idea into profit, selling his "Spalding's model" for $2.50. Soon the company was the leading sporting goods company in the country, being the first to make footballs, basketballs and golf balls (as well as golf clubs).

However the first of over 300 patents placed within patent classification 2/19, "baseball gloves", was US 290664 which was filed in 1883 by Austin Butts of Groveston, NJ. He explained in it that earlier gloves had four seams per finger but his only had two seams plus a pad that ran across the palm to the thumb. Only shortened fingers were provided as the fingers were meant to stick out from the glove. The next patent to appear in the classification was by a well-known name: George Rawlings, of St Louis with US 325968 in 1885. George and his brother Alfred opened a store selling sporting goods in St Louis in 1887. Using a thick glove or mitt took a long time to catch on, and the pitchers were the last to adopt them, as they thought

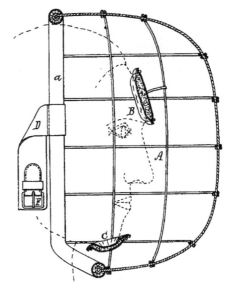

Catcher's mask (US 200358)

it would impede their pitching. Overhand pitching was only permitted from the mid 1880s which meant that hits became much harder. By the 20th century one commentator is supposed to have said, "The fielders have become so inseparably attached to their gloves that if you jerked one off a man's hand, you'd almost expect to see him bleed to death".

Rawlings Sporting Goods have continued to innovate with gloves. Harry Latina, the "glove doctor", was their chief designer in this area, with many patents published between 1925 and 1963. His inventions included the Trapper in 1941, a three-fingered, deep-well glove for first basemen, patented as US 2281315, and the Trap-Eze®, a popular six-fingered glove which still sells well. This invention was filed in 1958 and was published as US 2995756. Latina lived in East St Louis, IL, and the assignee was Spalding, who by then had taken over Rawlings. He explained that he wanted a versatile glove rather than different gloves for specific positions on the ball park. Other advantages spelt out were that the glove fitted more snugly round the wrist, padding was placed where it was really needed, and "the cost of constructing the same is greatly reduced and assembly thereof is simplified".

Turning to the ball itself, Benjamin Shibe of Bala, PA, applied for a patent for a "base-ball" in 1908 with US 924696. This was the first cork center ball, sheathed in rubber, such as has been used since 1931 (the cork center only being used earlier). Baseballs had originally consisted of india rubber with yarn wound tightly around it. Cork is less yielding and so the yarn can be even more compactly wound round it for a more "lively" ball. Shibe explained that the cork center lay within a

Baseball glove (US 2995756)

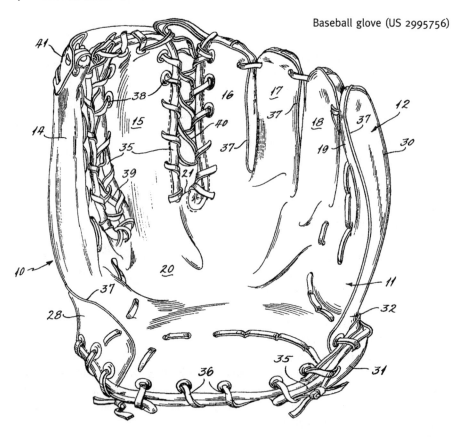

layer of rubber which was surrounded by yarn (preferably wool) which was itself surrounded by an "adhesive plastic" and then by stitched horse-hide or another suitable material.

During World War II there was a shortage of rubber. The Major Leagues temporarily used wholly balata cork centers rather than the rubber-cushioned cork centers to make what must have been a very minor saving. Lyman Briggs wrote about the resulting effect in a paper entitled "Methods for measuring the coefficient of restitution and the spin of a ball" in the January 1945 issue of the *Journal of Research* of the National Bureau of Standards (of which he was the director). He claimed that the ball helped the pitchers. "A hard-hit fly ball with a 1943 center", he wrote, "might be expected to fall about 30 feet shorter than the pre-war ball hit under the same conditions". The opposite effect may now be happening. Every baseball fan knows that the number of home runs has jumped in recent years, with the record for homers in a season being broken twice. In 1980 the average Major League game had 1.47 home runs but by 2000 this figure had risen to 2.34, 59% more. Various suggestions have been made to explain why this has happened, with one theory being that the balls have again changed. All Major League baseballs have been made by the same company, Rawlings, since 1976. This simplifies the task of analysing them.

Dennis Hilliard, a baseball fan (and also, incidentally, head of the University of Rhode Island's crime laboratory), tried dropping five baseballs made between 1963 and 2000 from a height of 15 feet. The three oldest balls bounced an average of 62 inches. The balls made in 1995 and 2000 bounced an average of 82 inches—32% higher. However, cork does begin to lose elasticity after about 10 years, and that may have altered their bounciness. Hilliard and his fellow investigators next tested the yarn windings. They found that the older balls' yarn was made almost entirely of pure wool. More recently made balls contained more and more synthetic fiber. Major League rules state that the yarn cannot be more than 15% synthetic. The yarn from the ball made in 2000 was more than 20% synthetic. This is important as wool easily absorbs moisture, and humidity can deaden a ball by relaxing the tension in the winding, and reducing the elasticity of its fibers. The two most used synthetic materials in the yarns are polyester and nylon—and both are resistant to moisture. Therefore the balls will indeed travel further.

The next of the big three is American football, which evolved from the English game of rugby. That sport came about when William Webb Ellis in a soccer game in 1823 at Rugby School suddenly picked up the ball and ran with it. More than any other popular sport, perhaps, the rules of American football were altered and experimented with in its early years. The game was rough from the start and it is very probable that the padding proposed in US 759833 by Walter Stall of Brockton, MA, filed in 1904, was needed. Instead of uniform padding it could be adapted to fit the individual with the padding being adjusted from the outside.

Helmets obviously protect the head, but the origins of the football helmet are controversial. A claim has been made for Joe Reeves, a player for the Naval Academy at Annapolis, MD. Reeves wore a helmet shaped like a beehive in the Navy's game against the Army in 1893 when his doctors feared another kick to his head would cause him to become insane. Other suggestions have been James Naismith (who earlier had invented basketball) in 1894 and George Barclay of Lafayette College at Easton, PA, in 1896, who wanted to prevent his getting cauliflower ears (as he liked the ladies).

In the 1930s close fitting leather helmets such as worn by early flyers, known as the "dog ear" type, became popular. Like earlier models they offered little protection for the ears. The first helmet looking similar to contemporary ones was by John Riddell Sr and his son John Jr, of Wheaton, IL, filed in 1940 as US 2293308. It was designed to be comfortable and light while still being strong. As so often in invention a new material was needed to give the right qualities, with 10 brands of plastic being suggested in the patent. The helmet was to be molded in one of these transparent plastics so that the painted decoration which (he states) was almost

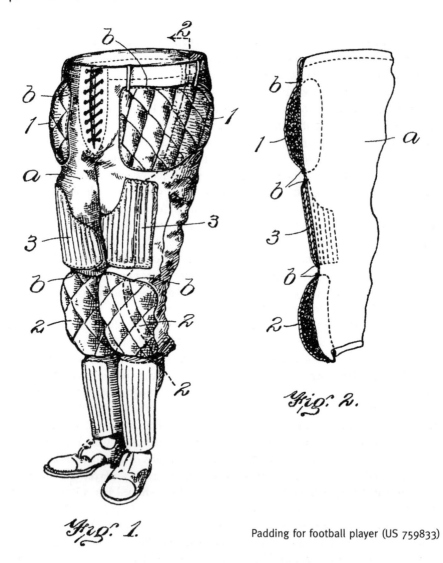

Fig. 1.

Padding for football player (US 759833)

universal by this time could be put on the *inside* of the helmet, to avoid its being scrapped off in normal play. A felt interior was then added to the inside.

The helmet was supposed to be made in two halves, with a reinforcing band (7) over the glued join. World War II's shortages meant that the invention could not be used for some time. Riddell Sr had taught maths at an Evanston high school while also coaching the football team. Among his other inventions he had earlier come up with the first removable football cleat (the first of 11 patents by him on the topic being in 1922, US 1602452), the profits from which enabled him to quit his job and form a sports equipment company.

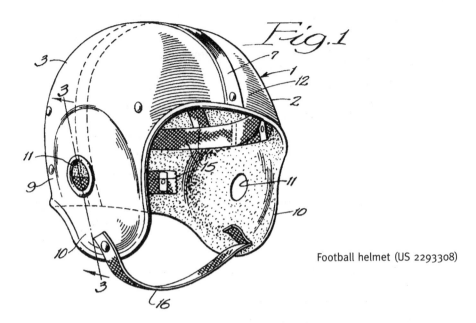

Football helmet (US 2293308)

Basketball was the third of America's major sports to be invented. The *Encyclopaedia Britannica* (Macropaedia, 1993) calls it "the only major sport strictly of US origin". Here again, as so often in the American Dream, a foreigner was responsible for what now seems wholly American. James Naismith was a Canadian working at the YMCA Training School at Springfield, MA, who later became a doctor and clergyman (as well as a US citizen). Dr Gulick asked him to provide a sport that the students could play indoors in the winter, when it was too late for football and too early for baseball. Gymnastics had been the norm but they were finding it boring. Naismith got his students to try playing football (players broke limbs), soccer (players broke windows), and lacrosse (players broke apparatus).

He discussed the problem with Dr Gulick who said to him, "All new things are simply a recombination of the factors of existing things". Naismith pondered what was needed. Contact between players was a problem on hard floors, so there would be no contact. Contact came from tackles, which came from running, so there would be no running. The act of throwing was softer than kicking a ball, especially if thrown overhand, so the ball would be thrown into a high box. He remembered a game he played when a boy in Ontario, "duck on a rock", where players had to throw stones to knock away a stone placed on a high rock.

Naismith asked the janitor, "Pop" Stebbins, if he had any 18-inch square wooden boxes handy. No, but he did have some peach baskets. These were nailed onto the base of the court's gallery, which just happened to be 10 feet off the floor, thus setting the regulation height. The date of the first game is uncertain but is thought to have been December 21, 1891, and was played with a soccer ball. Naismith's class were dubious when on entering the court they saw yet another sport, with Frank Mahan, a big Irishman who played on the football team, being remembered by Naismith for saying "Humpf, another new game". They had to be persuaded to leave at the end of the allotted hour. Naismith himself played center for one of the teams, which each had nine players (as he with the class numbered 18 in all). The peach baskets did not have a hole, and in the early games every time a player scored the game stopped while the janitor went up a ladder and poked a stick up a small hole in the bottom of the basket to get the ball out. Naismith later realised that it was easier simply to remove the bottom of the basket.

A problem with spectators sitting in a gallery was they often leant over either to help or prevent the ball from entering the net. The backboard was therefore introduced in 1895, originally being made of metal wire. Another aim was not to encourage rebounds, as now, but rather to cushion the force of the ball. However they were often distorted so as to influence the play and so by 1904 wooden backboards were required. Later the backboards were moved further from the end lines to reduce the number of players stepping out of bounds. Despite these changes US 878309, the first miniature game designed to simulate basketball, which dates from 1905, shows a mesh backboard or "deadening net". The inventors were John Pell and Edward Clark of Great Barrington, MA. Play was started by rolling the ball against the other end and the player who had rights to that square (as predetermined) tossed the ball (it is not explained exactly how) from whatever square it landed in. They talked of basketball being a "well recognized game".

The metal rimmed hoop replaced the basket from 1893 (but the first basketball as we know it only dates from 1909). The first breakaway hoop, to prevent harm if a player's arm caught the hoop while dunking a shot, was US 4348022 by John O'Donnell of Peoria, IL, in 1980. Another 38 have followed since in the same classification for a "breakaway hoop", 473/489.

One danger with previous attempts to solve the problem, O'Donnell points out, would be that "breakaway" might be taken only too literally, with the entire assembly falling onto players. There would also be a delay in the game while it was put back together again. His solution was to provide a plate (12) above and attached to the L-shaped mounting bracket for the hoop. The bracket included a "cylindrical energy-absorbing polymer mass" which was displaced if the hoop were tugged downwards (but not in normal contact by the ball). This meant that the "L" would swing downwards so that the hoop would rest against the backboard. A buzzer could also be attached to warn that a possible foul had occurred.

These three sports have very large numbers of supporters, but relatively few participants. Bowling is the exact opposite. Nine-pin bowling evolved in Germany in the Middle Ages. Dutch immigrants are thought to have introduced the game to America, and appropriately enough it was one of their descendants, Rip van

Miniature basketball game (US 878309)

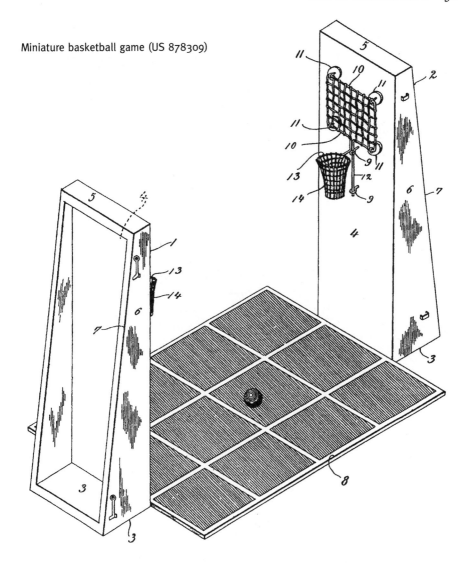

Winkle, who woke up in Washington Irving's famous 1820 story to the sound of "crashing ninepins". The sport became popular but, associated as it was with taverns, gambling became involved. In an effort to stop this, Connecticut banned "any nine-pin alley" with a state law in 1841, "An Act in addition to an Act entitled an Act concerning Crimes and Punishments" (this really is its name). Exceptions were made for those alleys where the local selectmen were satisfied that the alley was "used solely for the purposes of health and recreation". A fine of between $7 and $50 was the penalty for transgression. The familiar story is that the law was just too precise. It specifically outlawed *nine-pin* bowling. In fact the law did no

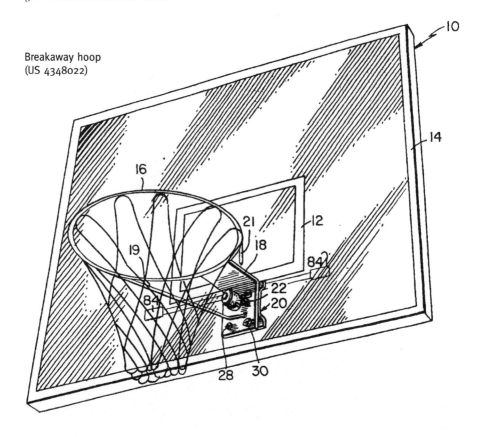

Breakaway hoop
(US 4348022)

such thing. It prohibited a "place for playing Bowles, skittles, or Nine-Pins, *whether more or less than nine-pins are used* in such play" (italics added). It shows how going back to the original sources, whether in patents or elsewhere, can sometimes demolish oft-repeated urban legends.

Nevertheless, it is certainly true that an unknown genius altered the game to the 10 pins, arranged in the familiar triangular formation, which is played today in the USA. Britain also uses 10 pins but Germany remains faithful to nine pins, arranged in a diamond formation. Indoor bowling alleys are thought to have first evolved in New York City in 1840 but the game was rather different from that played today. The reason is that we are used to the pins being set down by a machine. These only appeared in the late 1950s. Before then being a "pin setter" meant that you manually set up the pins. The key patent to end it seems to have been "Pinsetting machines" by George Montooth of Long Beach, CA, with his US 2817529, filed in 1947 but not published until 1957. Brunswick was the company assignee. It looks very complicated, and 14 pages of drawings and seven pages of description were needed to explain how it worked. It was for a new compact and integrated machine that would work within the usual confines so that it could be installed in existing alleys.

The mechanism was meant to cope with the normal two-ball cycle but would automatically adjust if the first ball were a strike or went in the gutters. The drawing shows the situation after a strike. The ball, having fallen onto rotatable rods in the pit behind the pins, rises at (A), the weight of the ball having caused the mechanism to raise the ball in its grip. The rods were inclined to one side so that at the right height the ball would topple into, and glide down, the ball return (109). Any pins that had fallen into the pit would be moved out of the way by the rotating rods. While (A) was at its maximum height the pin setting mechanism (E) would descend to pick up any pins still standing on their places before rising again. The guard and sweep at (F) then pushed the dislodged pins into the pit before returning to its normal place.

The mechanism (B) in the bottom of the pit then gathered the dislodged pins on the mat (141) which rose up and by means not shown in the drawing (but including piston (257)) took the pins to the rollers at the top of the drawing before dropping them into the long hopper (140). This was then raised by cable (149) and tilted so that the pins fell into the mechanism (D) which funnelled them down so that mechanism (E) could set the pins again in the usual formation. Handily, the heavier end of the "butt" part meant that the pins always came down the right way up. All this was in order to ensure a smoother, quicker game—and to save on the slender wages of the pinsetters. Montooth did say that his design would still make it possible for a "pin boy" to set the pins manually.

Such sports are thought of as old-fashioned by those who relish the spice of danger in their sports. "Extreme" sports have become very popular in recent years although they are easier to watch than to define. One possible definition is that they inspire fear in the participant or (increasingly) in the mind of the beholder, who perceives the skill but not the care being used to prevent injury. They are certainly spectacular to watch. Many such sports require much skill so that they cannot be casually taken up.

Dr Gulick's remark about inventions being combinations of existing things, which helped lead to basketball, is certainly true of extreme sports. Again and again sports are mixed up to create a new one, which might be more thrilling, or playable in the absence of, say, snow or of the ocean. For many this is truly what the American Dream is all about. Many such sports were first devised for fun, but growing audiences now appreciate them as spectator sports. The trouble, perhaps, is that there is such a bewildering variety of extreme sports that they are all vying for attention and for expert players. Many also require special facilities. As might be expected much innovation in the area comes from California.

For grownups, at least, the closest they are likely to come to an extreme sport (if it can be called that) is line-skating, which is simply a merger of the wheels of a roller skate with the single blade concept of the ice skate. As in ice skating it allows

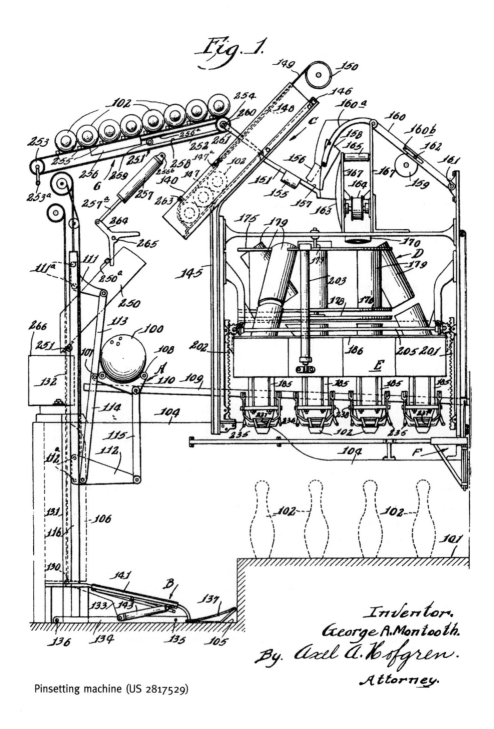

Fig. 1.

Inventor.
George A. Montooth.
By. Axel A. Hofgren.
Attorney.

Pinsetting machine (US 2817529)

much greater turns than do roller skates. The idea had been around for 150 years but never really got going. Scott Olson was a Minneapolis ice hockey player who had a couple of patents for ice skates. One day in 1979 he found a 1966 vintage Chicago inline skate in a used sporting goods store. By modern standards it was primitive but it did have four wheels in a row, with the front and back wheels extending beyond the boot. It would help him train in the summer, so Olson purchased the skates, and used his experience to modify them to include better wheels and a (rubber) heel brake, although he did not patent his changes. A new material, polyurethane, was needed to make it work. His company, Rollerblade Inc., was not the first company to manufacture inline skates but by offering a comfortable skate with a reliable and easy-to-implement brake it introduced inline skating to millions. In a way the company has been too closely identified with the sport as many talk of rollerblading, but Rollerblade is a trademark and not the name of the sport.

The displayed US 5028058 is an example of the company's inventions. It was filed in 1990 by Brennan Olson, Scott's brother. The patent explained that existing models were heavy, and their urethane wheels were liable to melt and come away from the hub after heavy use in hot weather. The upper drawing shows the new skate, mostly of synthetics, with side rails to house the axles, while the lower drawing shows in exploded form the new hub and wheel mounting structure. Each wheel was mounted slightly differently as each would meet different strains when in use. The brakes at the heel were replaced by a new design using synthetic materials, placed at (18) in the upper drawing.

For those who are younger, at least in heart, the skateboard may sound more attractive. The first marketing of the skateboard was by Bill and Mark Richards of Dana Point, CA, in 1958 together with Chicago Roller Skate Company. They attached roller skate wheels to a square board and sold them at their Val Surf Shop for $8 each. This seems to have been the first attempt to sell the product but many youths had built home-made skateboards in the 1930s and 1940s. The first skateboard contest was held at the Pier Avenue Junior School in Hermosa, CA, in 1963.

Over 300 American patents have been published on the subject, with the earliest to have the word actually in the title being US 3235282, filed in 1965 by Louis Bostick of Granada Hills, CA, entitled "Skate board provided with longitudinally adjustable wheel carriage units". These early skateboards were meant to keep contact with the ground, rather than trying out "aerials" or "verts". The story behind how such tricks came about goes back to 1975 and a group of teenage surfers called the Z-boys (after their favorite surf shop, the Zephyr Surf Shop) from "Dogtown", the run-down district between Santa Monica and Venice, CA. The movie *Dogtown and Z-boys* is a popular documentary about what happened.

The swell was much less in the afternoon, so instead of surfing they would sit around drinking beer and talking about their sport. The shop's owner, Craig Stecyk, suggested that they spend the time practising surfing moves instead—on skateboards. They tried crouching on the boards, as if they were surfing (rather

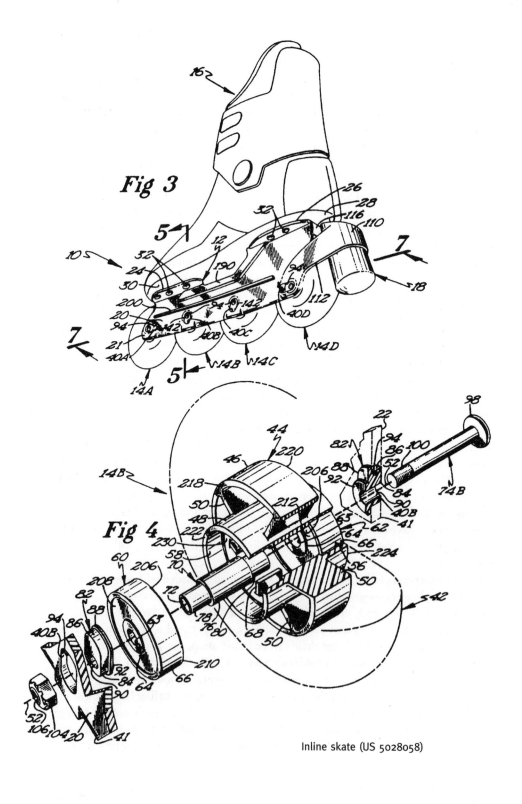

Inline skate (US 5028058)

than the usual standing position), and practised in playgrounds with sloping walls that they could ride like asphalt waves. Then drought hit Los Angeles, and many swimming pools emptied. The Z-boys started going through Beverly Hills standing on car roofs looking for empty pools with curved sides, in apparently unoccupied properties. The radio would be turned on, beers would be opened, and skateboarding began until the police appeared and chased them off.

So far they were still keeping contact with the surface while skateboarding, although the sides of the pools were much steeper than those they had previously used. Day by day the best skateboarders went higher and higher with their pivot turns. Then one day Tony Alva went too high and rose above the side of the pool. He instinctively grabbed the board's edge and turned in mid-air. This is thought to be the first "aerial". Steyck wrote up what they were doing in *Skateboarder* and publicity soon built up, with the first purpose-built bowl soon opening.

There has been an attempt to return to the original concept with Steen Strand's Freebord® design for a "Lateral sliding roller board". He is a designer and entrepreneur from San Francisco who filed in 1996 for US 5833252. One advantage is presumably that more people would feel inclined to use a safer-sounding, and behaving, board. These boards enable you to go sideways and not just forward and back, but you cannot do jumps.

The axles are much longer than the standard ones on skateboards and the "trucks" (wheels) have an in-built spring bias to help control by the user but not so much that the board is prevented from going sideways. This emulates the movement of snowboards. The action is performed by the two rollers, shown in the lower drawing as (50) and (51), which pivot to reflect the direction taken. Ingeniously, the rollers are of different heights so that the user can alter the movement by shifting stance and hence weight so that one roller will have no effect. Freeboarding has been compared to driving over an ice-covered pond, with lots of skidding and sliding, rather than the fireworks of the top skateboarders' acrobatics.

A team sport invented in California is the "Rollercross-rink design". It is explained in US 5906545, filed in 1998 by Robert Eden's Eden Enterprises. The invention is for the rink but gives a good idea of the rules along the way. The sport involves inline skaters playing lacrosse in a roller hockey arena with half-pipe modifications (that is, the ramps round the arena). The patent explains that ice rinks are both relatively scarce and time on them is limited by their popularity. Inline roller hockey is described as the fastest-growing team sport in America and is the closest you can get to ice hockey. The proposed sport emulates the style of that sport while using lacrosse sticks to hit the ball (which travels faster than in ice hockey) with halfpipes to add interest. "This game requires high-speed maneuvers, fast passing and dramatic free-form vertical skills thereby creating a visually exciting experience for the spectator and thrilling and competitive play for the participant". The name of the new sport is not given but trademark records show that at first Extreme Rollercross and then Rollercross® was used as the name, as

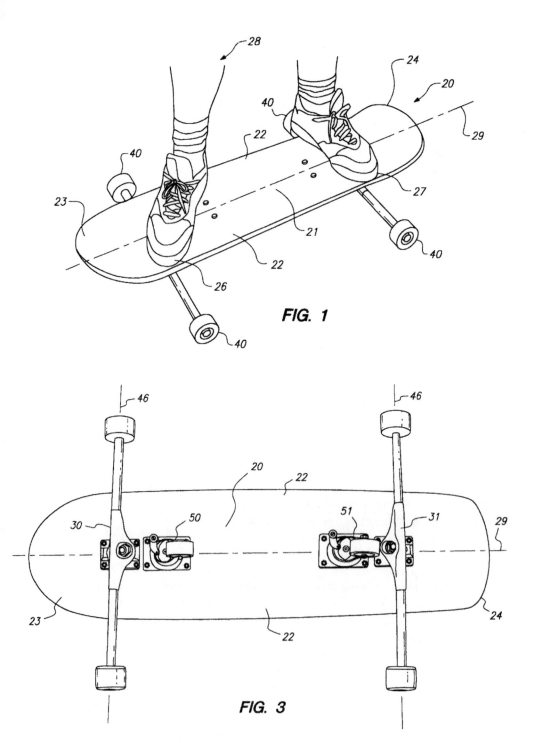

FIG. 1

FIG. 3

Lateral sliding roller board (US 5833252)

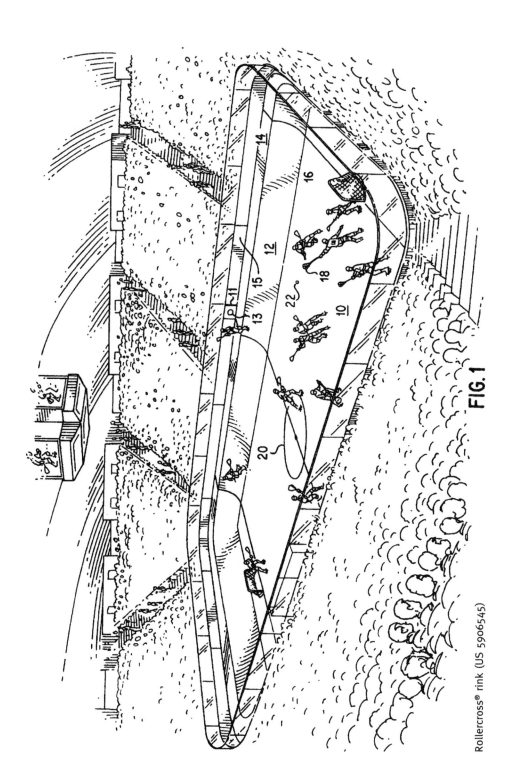

FIG. 1

Rollercross® rink (US 5906545)

suggested by the patent's title. The sport will, however, require new arenas to be built. It is in the hands of the inventors as to what future sports will evolve to challenge the current leaders.

That's entertainment

....

WHAT is the use of working if you cannot have a good time? A vast amount of time and money has been spent in entertaining Americans, mainly (but not entirely) in the hope of making still more money in exchange. Much of this entertainment is passively received, but some entertainment requires the active participation of the consumer. Each fresh generation seems to need fresh extremes of stimulation.

Disney is part of every child's life, and that includes their theme parks. The "imagineers" employed by theme parks are always thinking of fresh ways to entertain, excite and maybe terrify their visitors. Besides the effect of surprise, part of the fun is often not knowing how an effect is achieved. US 6309306 filed in 1999 by five Southern Californians for Disney Enterprises is an example of how to enhance the consumer's enjoyment. The invention uses the Treasure of the Incas® concept to explain its interactive, telepresence maze-exploration game system. This is an actual entertainment at the DisneyQuest Explore Zone at Orlando where miniature jeeps equipped with television cameras move around a real (not "virtual") maze to find treasure. When the treasure is found and retrieved by the jeep it appears on the user's console.

The inventors explain that visitors to such attractions are able to operate remote-control (RC) vehicles navigating an environment where sensors are tripped so that effects such as water, fire and lights are set off. "The guest cannot fully experience visually the acceleration, bumps, wakes, collisions, spin outs, or falls of the RC vehicles. Further, because the driver must be able to view the RC vehicle and the defined environment in order to navigate, navigation through a maze or night environment is not feasible because the driver could not see the vehicle through the walls of the maze or through the darkness. Thus, the driver is deprived of an entire class of experience, namely, exploration." Other problems included needing to have the maze itself in view so that the visitor could understand the context, and stopping all the vehicles navigating the maze if a single vehicle overturned. These were the problems that the patent set out to solve without going so far as to install each visitor in a real vehicle.

The concept is perhaps not quite as exciting in reality as it sounds, and resembles in many ways video arcade games. The old-fashioned hall of mirrors is used, but with some twists, as reality is mixed with video and special effects. Software runs the entire experience, with a "script" which is full of branches so that different things happen depending on what has occurred, or has been determined by the visitor, to avoid repetition. In order to enhance the feeling of speed, the patent recommends that the width of the corridors should not exceed four times the

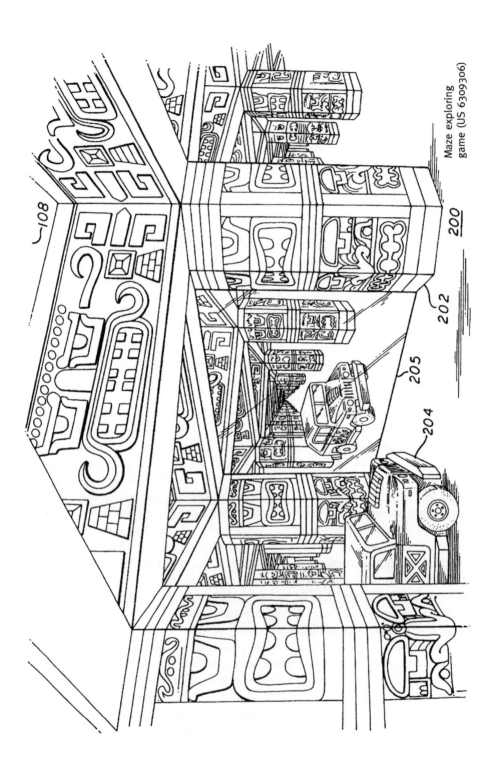

Maze exploring
game (US 6309306)

width of the vehicle. The maze is in the shape of a triangle rather than a square, and incorporates mirrors as some of its walls. Both of these help create effects such as the endless corridors like those shown in the drawing. Removable floor tiles mean that the damaged jeeps can be retrieved by the operator without stopping play for others. Video effects can be imposed on, or replace, the telepresence video signal. Hence the effect of two phantoms can suddenly appear to the startled visitor as the jeep races down the corridor.

A key part of the Dream has to be America's love affair with the movie industry. If there is one common thread it is the attempt to make the experience of watching moving images more vivid and realistic. More like real life, in other words. The origins of movies are complicated, but its beginnings arguably date back to an attempt to settle a bet following an argument between Californian capitalists.

Eadweard Muybridge was an English immigrant who ran a San Francisco bookstore until he began to get interested in the new field of photography, which he later made his profession. In 1872 former Governor and railway magnate Leland Stanford asked him if, as he believed was true, it could be shown that horses could have all four feet off the ground at once while trotting. Muybridge said that photography was not advanced enough to settle it. Stanford is said to have replied, "I think if you will give your attention to the subject, you will be able to do it, and I want you to try". Some stills were taken at the Sacramento racetrack and although one photograph seemed to be conclusive, the shutter speed was too slow to capture each motion and therefore the photographs could not be put together to show a sequence of actions.

Muybridge very nearly lost his chance of contributing to movie history. He discovered, through letters sent to his wife Flora, that she had a lover, "Colonel" Harry Larkyns, who was the father of the 6-month old whom he had supposed to be his own child. On October 17, 1874 Muybridge looked for Larkyns at a party at the Yellow Jacket quicksilver mines near San Francisco. He greeted Larkyns with "Good evening Major, my name is Muybridge and here is the answer to the letter you sent my wife". He then shot Larkyns, who expired on the spot. Muybridge was acquitted in his trial after an insanity plea by the defence, which referred to the after effects of a terrible stagecoach accident in 1860. The jury decided that Larkyns deserved what he got. Flora herself died a few months later.

Stanford went on to finance Muybridge's work with $40,000. On a warm June day in 1878 at Stanford's stud farm, in front of an audience of society people and newspapermen, his trainer rode a horse pulling a two-wheeled sulky cart in a trial. Electric circuits were completed by the wheels going over wires laid on the ground which set off 12 cameras in turn. These took pictures against a white cotton background marked with vertical lines which were numbered at 21-inch intervals. All 12 photographs were taken in less than half a second. The photographic plates were developed within 20 minutes and laid out on the grass, and finally showed

Photographing objects in motion
(US 212865)

that Stanford was right: a horse *can* have all four feet off the ground. A few weeks later Muybridge filed for his US 212864–5 which together showed how his "Method and apparatus for photographing objects in motion" worked. He specifically mentioned using it for horses, and an illustration in his patent represents a photograph taken in accordance with his specification (and the experiment).

The photos were widely published in America and Europe. *Scientific American* reproduced the photographs in its October 19, 1878 issue and invited readers to paste the pictures in the popular toy, still sold today, known as the zoetrope. This is an open drum with slits in its side, mounted horizontally on a spindle so that it can be twirled. As in a child's flicker book the phenomenon known as persistence of vision occurs: the still pictures, if seen rapidly enough, give the illusion of motion. This is of course the basis of moving pictures.

There is much controversy about exactly what innovations occurred when, and another book could easily be filled up by them. Many things came together: for example, celluloid was vital as it was the only material tough enough yet flexible enough to pass through film projectors. What gave impetus to the birth of the movie industry was (almost inevitably) Thomas Edison's team of workers, although he initially failed to appreciate the importance of what he regarded as a toy. He filed a "caveat" for the idea at the Patent Office in 1888 stating that he had an idea for an invention which would "do for the eye what the phonograph does for the ear". A caveat was a device to ensure that the Patent Office was obliged to warn him of similar ideas being filed.

In 1891 he filed a patent for the Kinetograph, his camera, which was not published until 1897 as US 589168. By then it was out of date but Edison hoped (unsuccessfully) to monopolize the business. The Kinetoscope, the viewing apparatus, was filed for in March 1893 but the patent was never published, presumably because it became even more out of date. This was in many ways an electrical version of the zoetrope and was housed in a cabinet. Film passed at 46 frames a second between a light bulb and the lens. Between the lens and a peephole was a spinning shutter with slits. The first peep-show parlor was opened on Broadway in April 1894 and caused a sensation. Little plot could be developed in brief movies such as *Buffalo Bill's shooting skill* and *Record of a sneeze* (which featured Edison's assistant Fred Ott, who could sneeze on cue), but audiences still flocked to see them. The trouble was that each person required one machine, and it would clearly be more profitable to pack everybody into a single theater.

Edison failed to file any patents abroad within the required 6 months' time limit so the field was wide open in Europe. The Lumière brothers in France, who owned the biggest photography factory in Europe, adapted and improved the technology for their Cinématographe in 1895. This device's name gave the world the word "cinema". Their idea was patented as French patent 245032, which was improved with four "additions". The patent covered both filming and, if adapted, projecting in the same camera. It was used for the first ever projection as we know it to an audience of more than one: *La sortie des ouvriers de l'usine Lumière*, a 48-second movie showing workers leaving their factory. This new technology soon spread and superseded Edison's work. The first ever feature movie with an actual plot, *The great train robbery*, was, however, made with a Kinetograph camera. This was as late as 1903, in New Jersey, by Edwin Porter. The final shot was, literally, of the bandit chief firing at the camera, which caused a sensation. The movie was one reel long—about 10 minutes.

New Jersey may seem a strange place for a Western, but Edison's research establishment at Orange was in the same state. The frequently cloudy and wet weather caused serious problems. Much filming was done on stages, which were open to the sky. This helped with the lighting, which needed to be very bright for early moviemaking. By World War I most of the industry had moved to the Los Angeles area, which besides being dry and sunny also had more varied scenery. The fightback against French movies, which by 1909 had 70% of the American market, was on. Foreign-language movies were not really at a disadvantage as long as movies continued to be silent. All that might need to be changed when showing abroad were the frames with dialogue or explanations. "Talkies" would have put foreign language movies at a great disadvantage, but before that happened the chaos of World War I meant that France and other European countries became minor players in the face of Hollywood's dominance.

Immigrants (or their children) were however heavily involved in the new industry, especially East Europeans. So long as they produced the goods the financiers were not bothered. Cartoons can be taken as an example of innovation, with a major pioneer being Max Fleischer (with his brother Dave). He was born in Vienna, Austria in 1884 and came to New York City with his family as a boy. As an artist and later animator he created Ko-Ko the clown ("Out of the inkwell", which gave the name to his studio). He also got permission from King Features to animate their Popeye character, although they did not understand why he was interested in something so ugly. Until his studio closed in 1942 it was the main competitor to Disney. In 1915 he filed from New York for US 1242674, the Rotoscope, which has been used by all the major studios for cartoons. It projected a frame at a time onto a drawing board where an artist could use the outlines for sketchwork. For example Dave Fleischer would clown around as Koko and the resulting film would then be used to draw the cartoon outlines. Sharp-eyed cartoon buffs have noticed in old Paramount cartoons, just below the famous mountain peak and stars, the inscription in tiny letters "Stereoptical process and apparatus patented. Patent number 2054414". This patent was filed in 1933 and enabled the

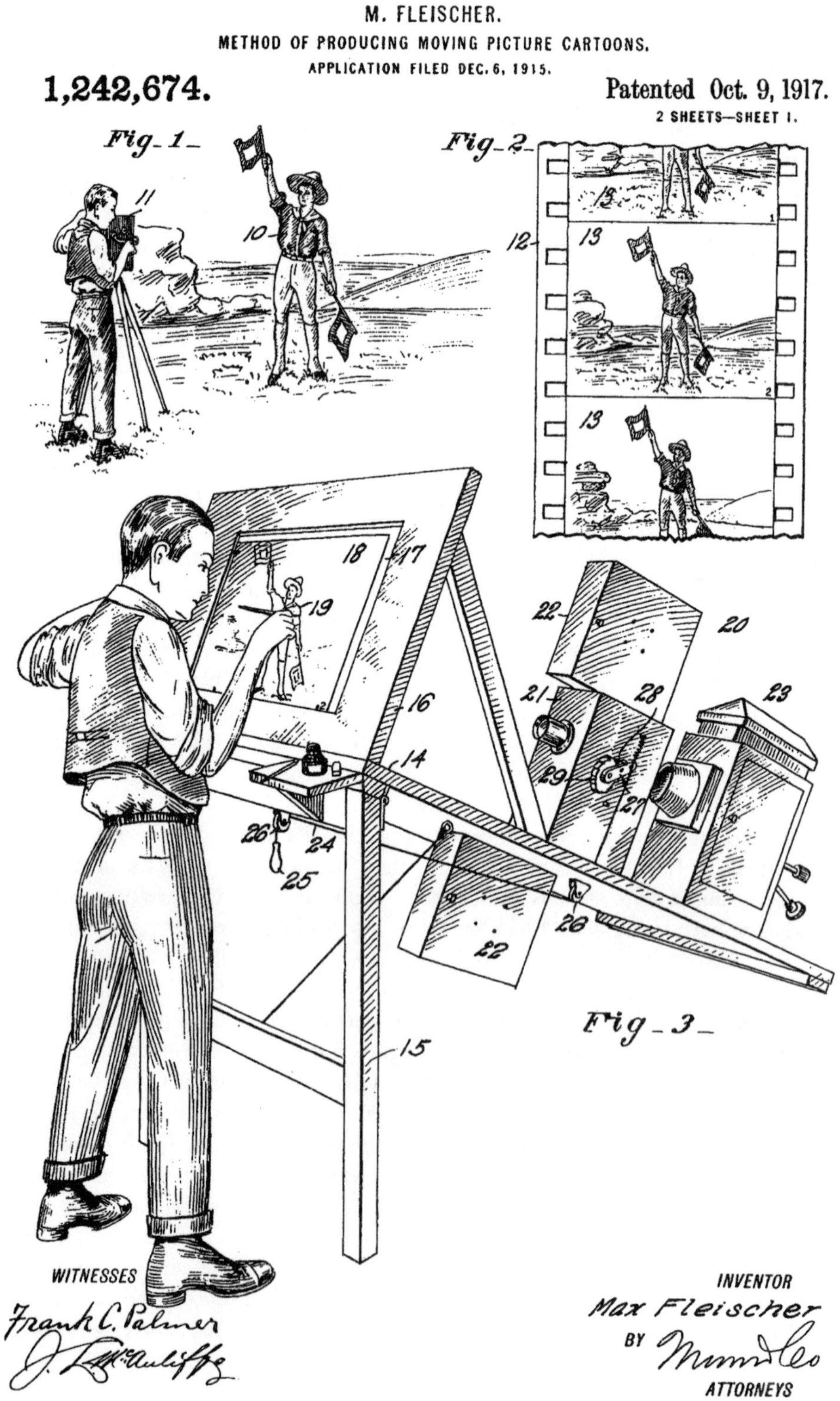

M. FLEISCHER.
METHOD OF PRODUCING MOVING PICTURE CARTOONS.
APPLICATION FILED DEC. 6, 1915.

1,242,674.

Patented Oct. 9, 1917.

2 SHEETS—SHEET I.

Fig. 1.

Fig. 2.

Fig. 3.

WITNESSES

Frank C. Palmer

J. W. Auliffe

INVENTOR

Max Fleischer

BY

Munn & Co

ATTORNEYS

Rotoscope (US 1242674)

surprisingly vivid three-dimensional backgrounds often seen in cartoons. If you look carefully at the main drawing (and enlarged below it) you can see Betty Boop, another of his creations, in action. Fleischer also secured a design for a Betty Boop doll, US D86224.

The invention was called the Rotograph. A three-dimensional set was built on top of a turntable. A camera was set up to film it as it rotated with a vertically-mounted clamp system holding the animation cels in front. Fleischer used the technique in such cartoons as the expensive color two-reeler *Popeye Meets Sinbad the Sailor* (1936), Popeye's first color appearance. More memorably, it was also used in *Who framed Roger Rabbit?* (1988).

Meanwhile sound had arrived with the famous words in *The Jazz singer* (1927), "You ain't heard nothin' yet!". When asked if his studio would produce "talkies", Harry Warner, head of Warner Brothers Studio, made the famous

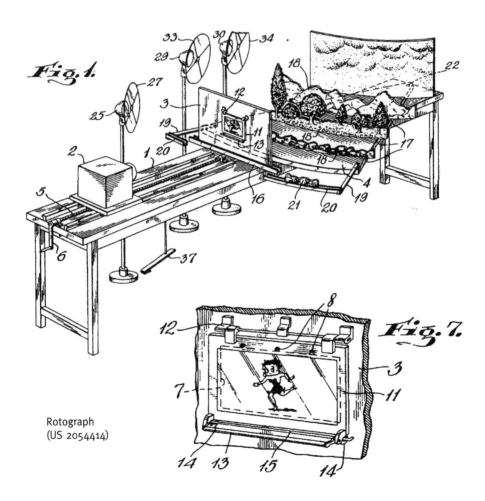

Rotograph
(US 2054414)

retort, "Who the hell wants to hear actors talk?" He turned out to be wrong. One little-mentioned effect of the "talkies" was that the 50,000 musicians who used to play the music in the movie theaters were suddenly out of work. Having dismissed the first all-sound cartoon as "a bit of racket and nothing else", Walt Disney from his Los Angeles studio several weeks later created his own first sound cartoon, *Steamboat Willie* (1928). This was the first appearance of the character who was later called Mickey Mouse. Mickey also appeared in one of his patents. Together with Wilfred Jackson and William Garrity, Disney filed in 1931 for US 1941341, "Method and apparatus for synchronizing photoplays". It was a technique to synchronize perfectly the sound with the appropriate images. The cartoon frames show in the one marked "175" a note being struck by the pianist which was to be followed by the sound of the next note being struck at frame 180. This was reflected in marks being made against the celluloid film so that a mechanical method would synchronize the sound with it. The patent suggested a normal speed of 24 frames per second, with the warning that it was important to plan the action in the frames so that the action would move in a "smooth and flowing manner".

Disney's studio also invented their own technique for placing animated characters within a three-dimensional scene, the Multiplane Camera. The National Inventors Hall of Fame™ cites US 2201689 as the patent but this is in fact for a different, if intriguing, technique. The Multiplane Camera is a misnomer as the concept involved painting the background on different sheets ("planes") of glass. A rigid framework supported each plane at the required distance. Cels with characters on them could be placed on any plane. The planes could then be manipulated and the result filmed to show the characters apparently strolling through a landscape. The far planes could be moved slowly with those in front moving at a quicker speed to mimic what the eye is used to seeing. The technique was complex and took up a large room, with several workers needed. It does not seem to have ever been patented (perhaps the idea was too obvious?) although Ub Iwerks is normally given the credit.

The Multiplane Camera was first used in the 1937 short *The Old Mill* and was used in most early Disney features. *Snow White and the seven dwarfs* (1937) was the first feature length movie to use it, and it was also used in scenes such as the circus train moving over the hills and through tunnels in *Dumbo* (1941), and the opening to *Bambi* (1942) with a tracking shot over the forest. The flight of the Darling children over London in *Peter Pan* (1953) was also partly a Multiplane shot. The Multiplane camera had advantages over the Rotograph as it was easier to line up and to predict the actions of characters over the background. It was, however, so complex that it is no longer used.

While all this innovation was going on, people had to have a place to watch the productions. The drive-in movie theater has become part of American legend due to its convenient combination of the movies and the automobile. Its inventor was

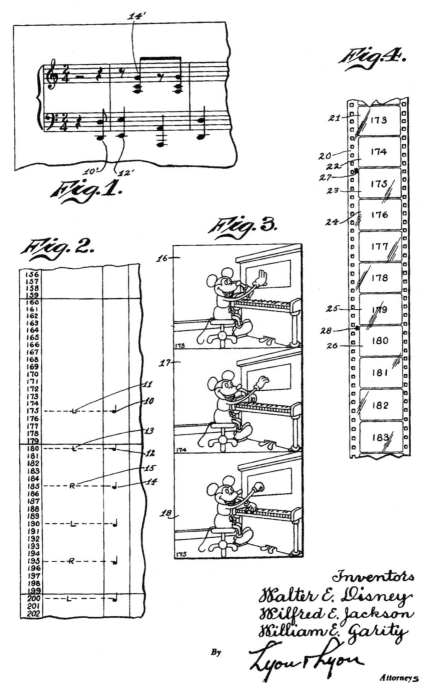

Synchronizing cartoons (US 1941341)

Richard Hollingshead of Riverton, NJ, who was a sales manager at his father's Whiz Auto Products. His mother told him that she hated the narrow seats in cinemas, and suggested that he do something about it. He wondered if sitting in a car might be more comfortable. He carried out experiments in the driveway by mounting a film projector on the hood of his car, and projecting movies onto a screen tied to trees. A radio was placed behind the screen for sound, and its quality was tested by leaving the windows closed, half open or fully open. Weather problems were assessed by using a lawn sprinkler to imitate rain. Hollingshead also thought out the problems of parking the cars. He quickly realized that cars could block the view for those behind, but by spacing cars at set distances, and placing ramps for the front wheels, good visibility was ensured.

The illustration from Hollingshead's US 1909537, filed in 1932, shows the idea. It was "a new and useful outdoor theater and it relates more particularly to a novel construction in outdoor theaters whereby the transportation facilities to and from the theater are made to constitute an element of the seating facilities of the theater". After passing the admission booth (25) the car would go to one of the parking spaces which sloped down towards the screen. Hollingshead suggested that by carefully using the inclines the drivers could avoid using their engines to enter or leave their "stalls", which sounds rather hazardous. Cars could also be parked beside the screen for those who preferred to sit outside and risk the weather. The projection booth was (26). Figure 5 shows the funnel around the projector. Ingeniously, a nozzle (31) fed in air to discourage insects from entering the funnel, which minimized the chance of giant insects being projected onto the screen. Providing that the land was available, the costs to the owner were minimal, with few seats and no roof or walls to build or worry about.

On June 6, 1933, in the midst of the Great Depression, Hollingshead opened the world's first drive-in movie in Camden, NJ, with an investment of $30,000. It cost 25 cents per car plus 25 cents per person. It took 10 months for another to open, in Pennsylvania, but by 1948 there were over 800 across the country. At first the idea of a drive-in movie theater was strange to people. To help them understand how it all worked, new drive-ins hosted "Open House" during the day where people were shown how to park, how the sound system worked and what food was available.

The sound quality at first was poor as it relied on bullhorns attached to the screen, which must have been noisy for the neighbors. Later on, in-car speakers greatly increased drive-ins' popularity so that their numbers grew to a peak of over 4,000 in 1958. At one time a third of all movie theaters were drive-ins. There was more privacy than in the usual cinema, especially when the car's occupants felt romantic, and they avoided the heat of the indoor theaters on hot summer evenings. Some drive-ins had rides as well, and concessions selling food enjoyed excellent business. Often a button could be pushed on the sound system to ask for food to be bought to your car. Then the market began to decline because of the

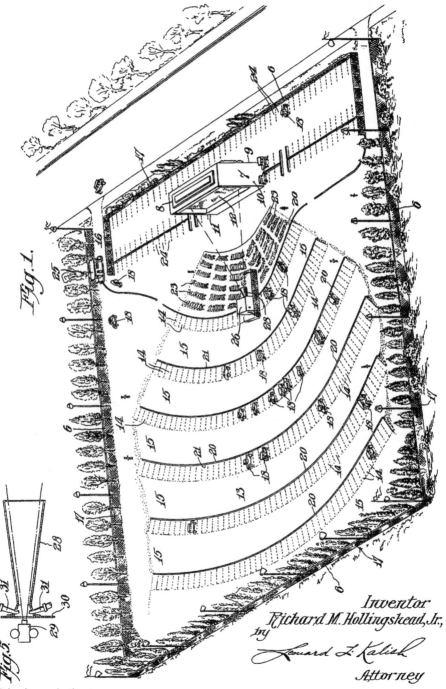

Inventor
Richard M. Hollingshead, Jr.,
by
Leonard L. Kalish
Attorney

Drive-in movie theater
(US 1909537)

impact of television, multiplexes and more profitable uses for land so that now there are fewer than a thousand drive-ins.

Even while the drive-ins were booming, the menace of television was drawing ever nearer. Much research into its practicalities had taken place over the years, with the first mention of "television" in the title of an American patent being in 1923, by a New Yorker, Alexander McLean Nicolson, who was stated to be a British citizen. In patents published in 1930 there were 23 such mentions; in 1940, no fewer than 155. It was realized that the concept was sound: it was just the small matter of getting everything to work properly by sending a signal which could be picked up and interpreted correctly in each home.

Among many major pioneers perhaps the most important was Philo Farnsworth. He was born in 1906 in what is now Manderfield, UT, but grew up on his father's farm near Rigby, ID. At the age of 14, while mowing his father's hay field, Farnsworth had a vision of images formed by an electronic beam scanning a picture in horizontal lines just like the mown field before him. He studied the subject and thought of adapting two known devices, the photocell and the cathode ray tube, into a single system. He showed his design to his chemistry teacher who many years later gave evidence of Farnsworth's "priority" in a court case with RCA, who had to agree in 1939 to license the technology. One day Farnsworth applied for a job with the Salt Lake City Community Chest campaign. Investors Leslie Gorrell and George Everson, who were conducting the drive, hired the young enthusiast who kept on talking about the potential of television. Everson agreed to finance his idea. More money was required to finance the research work but in October 1926 a laboratory was set up in San Francisco "to take all the moving parts out of television". Farnsworth filed his US 1773980 in January 1927 which is usually considered the first workable electronic (as opposed to the original dead-end mechanical) television. Farnsworth was just 20 years old. The basic idea was that an "image dissector" was used to scan the image for transmission. At the receiver, an "oscillite" tube reproduced the picture. Farnsworth's first successful transmission was of a dollar sign, with 60 lines making up the image.

World War II held up the launching of a television service, although radar ensured that there were going to be plenty of ex-servicemen trained in the basics of electronics to repair them. Radar also helped the development of the picture tube. The production of televisions began in 1946. At first there must have been a chicken and egg problem: there was no point in setting up stations if no one had a TV, and no point in buying one if there were nothing to watch. Standards were also needed to ensure that the sets could pick up signals from many stations. A meeting of industrialists as early as 1941 adopted a standard of 60 frames per second and 525 horizontal lines across the screen (this is still used today). NBC and CBS were radio networks which moved their radio talent (along with enthusiastic advertisers) to the new medium. From 1953 ABC, which had previously lacked both capital and affiliates, was also able to compete with them as a network. Du Mont

Laboratories, an actual manufacturer of the sets, struggled for a while as a network but as it did not have the radio talent to draw on it ceased broadcasting in 1956.

In 1948 there were 345,000 TV sets, each priced at several hundred dollars. By 1952 there were 17 million sets and, by 1991, 93 million households had at least one set. The first color electronic sets were manufactured in 1954. They cost $1,000 and there was little color broadcasting to receive. It was only from 1964 that more than 1 million color sets were made annually.

An example of a radio show moving to television was the big hit *The Howdy Doody Show*, which featured a marionette operated by the host, "Buffalo Bill"(Bob Smith), with his catchphrase "It's Howdy Doody time !". It originated with a radio show (which must have left much to the imagination) on NBC New York radio affiliate WEAF in 1947, *The Triple B Ranch*. The three Bs stood for Big Brother Bob Smith, who developed the country bumpkin voice of a ranch hand and greeted the radio audience with, "Oh, ho, ho, howdy doody". Martin Stone, Smith's agent, suggested to NBC that the show could transfer to television. The new *Puppet Playhouse* began on December 17, 1947 but within a week the name of the program was changed to *The Howdy Doody Show*. Children loved the Doodyville inhabitants because they represented American icons. The original marionette was designed by Frank Paris, and in keeping with Smith's voice was a country bumpkin. In a dispute over licensing rights Paris left the show with his puppet. The new-look Howdy Doody first appeared in March 1948. He was an all-American boy with red hair, supposedly 48 freckles (one for each state in the Union), and a permanent smile.

Smith treated the marionettes as if they were real and so did the children in his audience. In *Howdy and Me* Smith noted "Howdy, Mr. Bluster, Dilly, and the Flub-a-Dub gave the impression that they could cut their strings, saunter off the stage, and do as they pleased". A design patent for Howdy Doody, US D156687, was filed in 1949 by Robert Allen of Pacific Palisades and Melvin Shaw of Beverly Hills, CA, on behalf of NBC and will bring back nostalgic memories to many millions. Sadly, the show ended in 1960 after 2,343 programs. Each live-action show had required several marionette operators, and cartoons had no such costs.

An important innovation in the infant television industry was the remote control. Robert Adler of Northfield, IL, an Austrian immigrant, invented the first workable TV remote control, a great boon for couch potatoes. As the number of sets soared, Zenith's head, Eugene McDonald Jr, had thought that viewers would appreciate being able to avoid commercials. The first remote was produced in 1950, and was dubbed "Lazy Bones". It certainly enabled the viewer to turn the TV on and off, and to switch channels, but as it was connected by a cable to the TV it was obviously dangerous whenever anyone headed off to the kitchen to replenish supplies. In 1955 Zenith produced the Flashmatic, a device which was pointed at photocells at the corners of the TV cabinet. The problem here was that the photocells reacted to sunlight as well as to the beams of light flashed at them.

Fig.1. Fig.2.

It's Howdy Doody time! (US D156687)

Adler was the Zenith engineer who came up with a solution, with US 2817025 in 1956. The new remote communicated with the set by using ultrasonic sounds (which humans cannot hear) rather than light. Pressing buttons depressed one of four aluminum rods. Each rod emitted different sounds, and the television would interpret each of these as channel-up or down; sound up or down; and power on or off. The drawing shows the button (206) which on being squeezed depresses rod (207) down towards the spring (210) which enables it to rise up again. The workings were contained within the "vibrator element" casing shown as (201), with each of the four rods differing in the length of the vibrator elements. A miniature hammer (220) struck home to create a sound. Most of the patent explains the electrical circuits needed to interpret the sounds correctly. Unfortunately the cost of incorporating the technology meant an initial 30% rise in the cost of the televisions, but consumers soon warmed to the idea. After all, as the patent stated, "It is highly desirable to provide a system to regulate the receiver operation without requiring the observer to leave the normal viewing position". In the 1960s the system was modified by Adler to respond to electronic signals.

All this activity in the television business gave a simple alternative to going out to the movies to be entertained. Annual movie attendances had peaked at 90 million in 1930. The impact of the Depression meant that they fell to 60 million by 1932. The numbers had just recovered to their old peak in the late 1940s when the impact of TV began to hit and by 1953 sales had almost halved, to 46 million. It is little wonder that the movie industry panicked. At first they refused to sell movies to television, which was hungry for material, but from 1954 pre-1948 movies (which did not require "residuals", extra fees to actors and technicians) were sold, and the movie studios soon began making episodes specially for television. Nowadays, of course, the massive amount of network and cable television programming is in turn challenged by the lure of both computer games and by the internet.

Television was using the same 1 to 1.33 ratio on the screen as the movies had done. An obvious response by the movie studios was to make their productions even more vivid by making, literally, bigger movies. Cinerama was the first of many such attempts to fight back against television. Fred Waller and Richard Babish improved and simplified with patents such as US 2598731, "Diaphragm control for multilens cameras", a system which had originally been shown at the 1939 New York World Fair. Their new version was released in 1952 in New York with a special movie, *This is Cinerama*, which featured thrills such as a roller coaster ride and an airplane flight over Grand Canyon, accompanied by multi-track stereo sound. It ran for over 2 years. A curved screen was needed and the filming was carried out by three cameras at set intervals and angles. Projection was done in an identical layout so that the three images blended to give a rich illusion of space. However incorrect synchronization would mean that the lines between the images became visible, and each theater had to go to the trouble and expense of spending

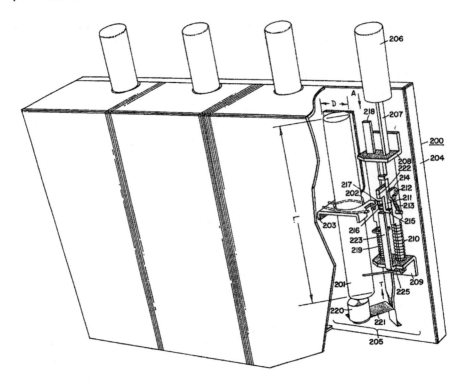

Remote control (US 2817025)

$75,000 for the necessary equipment (as well as three projectionists). *How the West was won* (1962) was the best known film to be made in this format.

Mike Todd, the producer, was instrumental in pushing for Cinerama but fell out with his partners before it first appeared. He had come to think that it was not good enough. He asked around for the "Einstein of optics" and was directed to academic Brian O'Brien, the President of the Optical Society of America. O'Brien was baffled at receiving a phone call from a man he had never heard of asking about a movie system of which, again, he had never heard. He eventually agreed to meet Todd at a bar across from the airport at the city where he lived, Rochester, NY. Todd arrived in a chartered plane and explained the problem to O'Brien and his assistant Walter Siegmund. "What I want is Cinerama out of one hole. Can you do it?", he concluded.

O'Brien said that it might be possible, declined the immediate offer of a job, but suggested that American Optical, the company he was about to move to as Vice President of Research, might be able to help. O'Brien also sent Siegmund down to New York City to see Cinerama for himself. Although able to see the movie with

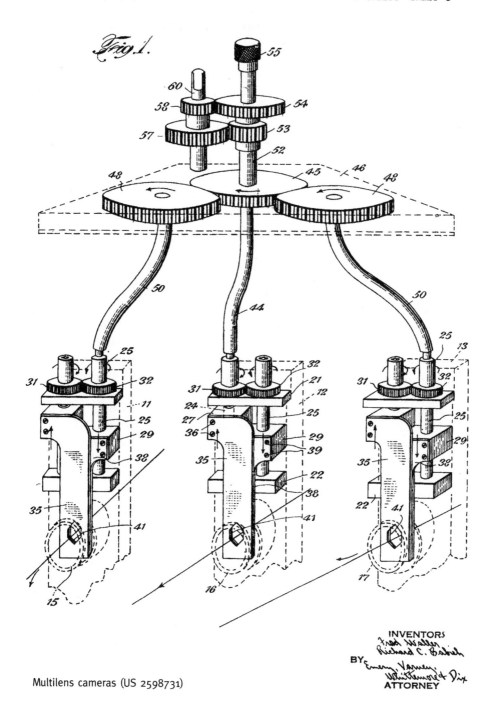

Fig.1.

Multilens cameras (US 2598731)

INVENTORS
Fred Waller
Richard C. Babish
BY Emery Varney,
Whittemore & Dix
ATTORNEY

expert eyes, Siegmund was still knocked out by the impact. The brightness was unusual because three images were being used, not just one. He reported back, "Wow!". Todd distrusted big companies but finally agreed to meet American Optical's President for lunch. After shaking hands he handed over a check for $60,000 and said "Let's talk business", which had the desired effect.

A new company called Todd-AO was formed. The first decision was to use a large film format so that there was increased brightness. 65 mm was chosen for the film format as some old cameras were found which used that size. Six audio sound tracks were added beyond the sprocket holes, to make the format 70 mm. Placing the sound track on the film avoided synchronization problems. Instead of three cameras there would be one, with a family of four lenses used to give special effects. Robert Hopkins, an expert in wide-angle work, who had succeeded O'Brien in his old University of Rochester job, was hired to design the lenses. Two were for close-ups: a large one for wide-angle, and a 128 degree "bug-eye" for very, very wide angles. "That lens today is almost trivial to design, but it was not trivial when you had to do things on desk calculators", Hopkins recalled. The result was Hopkins' US 2803997, filed in 1954. The biggest "bug-eye" lens was a massive 9 inches across.

It is probable that only a few experts can appreciate the details of the patent, but all can see the finished product. The first movie to use the technique was *Oklahoma!* with the "bug-eye" being used very selectively, such as the opening shot in the field of corn, and a runaway buggy. Todd hedged his bets by filming in standard 35 mm as well as his new system. A problem was that wide-angle lenses gave distortion. O'Brien and his son together with Siegmund worked on techniques to overcome this and suggested a complex technique of processing the film. The result was scratches on the film. The team had not expected this, as they did not know how movies were processed. After desperate repair attempts it was fixed for the first screening. The slogan was "You're in the show with Todd-AO".

The future of Todd-AO looked good, and the same system was used for *Around the world in eighty days* (1956) and *South Pacific* (1958). Tragically, Todd was killed in an airplane crash in 1958. American Optical lost money on their venture as the technology was superseded by Twentieth-Century Fox's CinemaScope, an inferior (but cheaper and simpler) process that originated from French technology. It used the anamorphic principle of squeezing an image in filming and then widening it when projecting so that the audience saw a very wide (but not "wrapped around" as in Cinerama) image. The first movie made with CinemaScope was *The Robe* (1953). By 1957, 85% of all US theaters were equipped for its "shoebox" screen format. Less successful systems included Cinemiracle, Thrillerama, Wonderama, Circarama, Quadravision and VistaVision. Each seemed to require its own filming and projection equipment, with the need for special glasses having already killed off a short-lived "3-D" craze, and it is little wonder that most were swiftly abandoned. The major problem, besides the cost of equipment, was always the distortion which widened anything on the screen. By 1967 CinemaScope too was abandoned, but it has left us with the now familiar 2.35 to 1 ratio between width and height rather than the old 1.33 to 1 ratio from older movies, and is the reason why many movies have to be seen "letter-box" fashion on TV to avoid losing some of the picture.

Movie formats such as CinemaScope and Todd-AO had multiple sound channels which were recorded on stripes of magnetic material applied to each release print. To play these prints projectors were fitted with playback heads like those on a tape recorder, and cinemas were equipped with additional amplifiers and speaker systems. From the beginning such formats had at least one channel played over speakers near the back of the theater. At first this was known as the effects channel and was used only occasionally for dramatic effect such as voices in religious epics. The effects channel came to convey greater sonic realism overall, not just the occasional dramatic effect. This expanded, more naturalistic practice came to be known as "surround sound", and the effects channel as the "surround channel". The extra speakers at the rear, and now along the sides of the theater as well, create a more diffuse sound field, which came to be known within the industry as "the surrounds".

Virtual reality is a very exciting concept, with the player being immersed, and interacting, with an entirely new environment. Typically a helmet is worn so as fully to envelop the senses. The first such devices were invented by Morton Heilig, a Long Beach, NY, movie maker. He dreamt of replacing the normal 2-D picture, taking up say 18% of the field of vision, with a 3-D experience taking up 100% of the field of vision. In 1957 he filed for US 2955156, which was for pairs of goggles with each eye having a colour television screen. Earphones were to provide stereo sound while nozzles provided air of different speeds (to suggest driving) or scents. This was adapted and he filed for US 3050870 in 1961. The patent had the title "Sensorama Simulator".

Heilig had moved away from anything like helmets towards what is rather closer to a video arcade game. Again there were to be breezes and scents, and vibrations could be created to "simulate actual impacts". It is clear from the patent that he envisaged using the concept not for amusement but rather for training in dangerous practices such as flying, or for training the armed forces. Students would learn quicker "with the least possible danger to their lives and to possible damage to costly equipment". Nowadays of course such training is provided by flight simulators even if they lack the breezes and scents. The problem was that the computer power needed was simply not available, or at least not cheaply, at the time, and so Heilig failed to get backing for his idea.

Instead work began on more modest video games. The origin of the industry is somewhat controversial, and while Nolan Bushnell established the arcade video game industry, Ralph Baer deserves the credit for inventing the video game with his US 3728480, filed in 1968 from New Hampshire for Sanders Associates. It talks

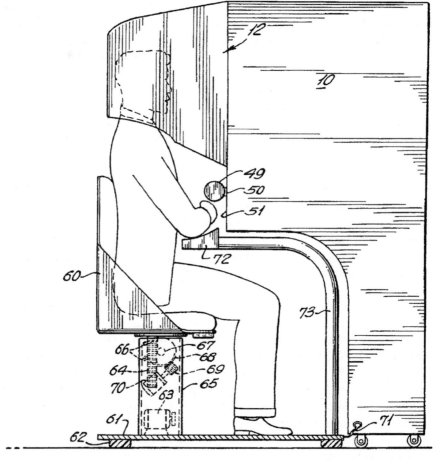

Sensorama simulator (US 3050870)

of a gun with a photo-electric cell which can be aimed at the screen to hit a dot which will then register as a hit. This is the basis of most modern arcade "shoot-'hem-up" games, no matter how they are dressed up. A 1976 court case in Chicago established that the Baer patent had priority over Bushnell's work.

Turning to music, it is impossible to say who invented the jukebox as the concept of inserting coins to play music has gradually evolved through many different models. By the 1930s the standard techniques had evolved, but looks as well as technology were vital. This is shown by the classic Wurlitzer 1015 model which was the subject of a design patent by Paul Fuller of Buffalo, NY, in 1946 for the company (US D146175). He called it a "phonograph cabinet". It was nicknamed "The Bubbler" for the effect of (heated methylene chloride) bubbles constantly moving up between the brightly lit glass arches at the top, which were programed to change color at intervals. It was an idea which they had been working on since 1938, when a champagne sign with a similar effect was shown to Fuller. Its looks helped sell 56,000 in the first year of production. This, more than any other, was the jukebox to which so many danced to rock and roll since, after all, the company's product was marketed under the slogan "Wurlitzer means music to millions".

Musical instruments cover a huge range with perhaps the most popular now being not the piano, as in the old days, but rather the guitar. It may be controversial but Clarence "Leo" Fender is often credited as the inventor of the first workable electric guitar, and certainly of the first popular model. An electric guitar has a solid body and does not amplify the volume of the instrument as do acoustic guitars. Instead it has electromagnetic "pickups" that convert the vibration of the steel strings into electric signals which are fed to an amplifier through a cable or radio device. Since it does not need to be naturally loud the body of an electric guitar can be of virtually any shape. With Clayton Kauffman, who was also of Fullerton, CA, Fender filed in 1944 for US 2455575, "Pickup unit for instruments".

The guitar has an odd shape with the neck and body being bolted together without glue or other fancy bits to make a cheap yet durable body. This set a fashion. Fender had 32 patents in all. The last to be published, US 3686993 in 1972, was called "Shoulder strap-operated pitch-changing means for Spanish guitars". His patent drawings are as attractive as this title. Half the fun with a Gibson guitar must have been names such as the Broadcaster, the Telecaster and the Fender Stratocaster.

Wurlitzer Bubbler (US D146175)

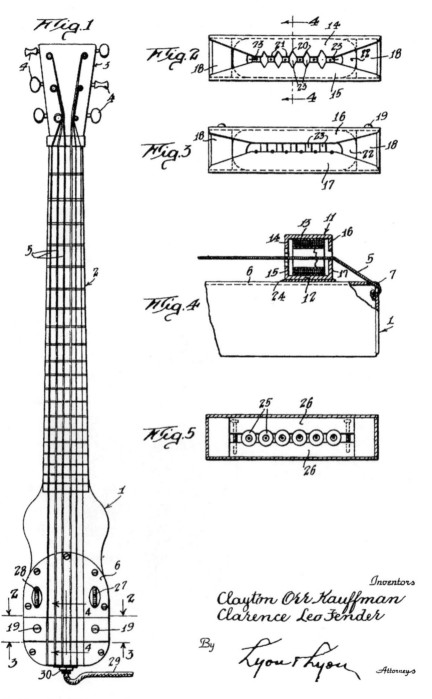

Electric guitar (US 2455575)

No significant advances have been made in the design of the electric guitar since the late 1940s.

The well known musicians Alex and Edward "Eddie" van Halen were born in The Netherlands and moved to Pasadena, CA, as children in 1962. They were trained as concert pianists by their clarinetist father but then their interests shifted, Eddie to drums and Alex to the guitar. Eddie had a paper route to pay for his drum set, so while he was out delivering, Alex played his drums. Alex became better than Eddie so he stayed with the drums while Eddie started playing Alex's guitar. They formed a band, Van Halen, although later Eddie went solo. In 1985 he filed from Los Angeles for US 4656917. This invention, he claimed, "leaves both hands of the player free to explore the strings which overlie the guitar body and fretted neck". It does this by a folding plate which swings out (by gravity) from the back of the guitar when it is picked up. It is then locked into place by a spring and the guitar is supported on the upper legs, with both hands free "to create new techniques and sounds previously unknown to any player". A drawing in the patent shows a musician in full flow using the invention.

Traditionally this music was made first on singles or LPs, then on cassettes and now on compact disks. MPEG Audio Layer 3, or MP3 for short, was adopted as a standard for transmitting music over telephone lines by a subgroup of the International Standardization Organization (ISO) in 1992. In 1996, Pentium-class PCs with massive hard disks provided the first suitable PC for high-quality audio compression (which is what MP3 does). The internet was rapidly gaining in popularity and with MP3 massive downloading of music could occur over what was, after all, just another telephone line.

The technology was based on German work by Fraunhofer Gesellschaft and was included in a family of patents, particularly US 5579430, "Digital encoding process", originally filed in 1989 and published in November 1996. By then the technology was well established on the internet as the ISO had released the source code for the MP3 software. In September 1998 Fraunhofer sent letters to MP3 software developers asking for royalties. Work suddenly stopped and free sites on the web began to disappear. Attempts have been made to design around the technology in the patents to avoid infringement. There is a dilemma: the music industry insists that it is losing money and that music will stop being produced if the artists and their producers do not make money, while others see a restriction of personal freedoms and feel that, after all, it is great publicity for the artists who will see many new fans at their (priced) concerts. Here is a profound divide between profits (or just making a living) on the one hand, and of enjoying music at a minimal cost on the other.

There's no place like home

EVERYONE wants to own his or her own home. The Department of Housing and Urban Development affirms this, with its website saying that on it, "You will find information about how you, too, can live the American Dream". For many this is about having a big house in its own lot, with a neat front lawn and a messier yard in the back for the children to play in. In 1999, 66% of all households owned their own homes. This figure is much higher than in most of Europe, where many rent houses, although it is comparable with Britain. The sheer size of America makes ownership a believable dream for many, although plenty still live in cramped row houses or in apartments. This is admittedly sometimes from choice: not everybody wants to live in the suburbs.

Rebels against Pete Seeger's "little boxes" have always existed. There is no particular reason why houses have to be shaped with lots of right angles, particularly when there are new ideas in materials and design. There have always been those who thought differently, such as Harriet Irwin of Charlotte, NC, who applied in 1869 for US 94116. This was a hexagonal house containing three hexagonal rooms and three lozenge-shaped rooms. The patent says that it was an economy measure, but the story goes that she hated cleaning dirt out of corners. So there were not any traditional 90 degree angles. Another unusual house was by James Lafferty of Philadelphia with US 268503, filed in 1882. The rooms are within the body of a massive elephant, with an "observatory" in the howdah on the top. The stairs are within the legs while the trunk descends to a symbolic barrel placed on the ground and provides water and drainage.

A more substantial innovator in housing design (and in much else) was architect and engineer Richard Buckminster Fuller. He was far ahead of his time in trying to conserve resources to avoid ecological degradation. He saw no reason why houses could not be mass-produced in factories, with standard parts being quickly erected on-site, rather than every house being custom-built. There had previously been such attempts, but based on conventional-looking houses. Dome-like structures were a source of special fascination to him, as he realized that they covered the maximum space with the minimum amount of materials, and were self-bracing. In 1941 he filed for US 2343764, the "Dymaxion house". His word stood for DYnamic MAXimum tensION and the invention used tension suspension from a central column to support the structure.

The drawing shows small windows but they could be much bigger without damaging the strength of the structure. The house was to be built of galvanized steel, with underfloor heating provided by the hot water pipes running under a conducting-material floor. It had several advantages over a normal house. It could

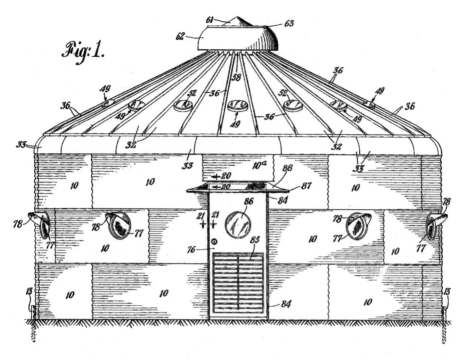

Fig: 1.

Dymaxion house (US 2343764)

quickly be assembled; rooms could easily be adjusted in size by moving interior walls; it was resistant to storms and earthquakes; and little maintenance was required. It weighed 3,000 lb in contrast to typical houses weighing 150 tons. Ventilation was from above and blew dust into ground-level filters. To save space the closets rotated, just as in many modern kitchens.

The layout inside the houses was determined by dismountable walls of plywood which were connected by hinges. Doorways leading off a central hallway allowed for entry to the other rooms. There would have been wasted space between the walls and the beds or couches, always a problem with curved walls, although this problem would be reduced in larger circular structures. Sound insulation might also have been a problem. The only house to be built to Fuller's specification was constructed in Wichita, KS, in 1948, and was inhabited for decades by one family. Beech Aircraft were persuaded to build such houses at their aircraft factory as the materials used in both the Dymaxion houses and their airplanes were similar. Unfortunately the project never went into production, as suggestions by the company which would have been compromises were not accepted by Fuller. In 1992 the house was carefully dismantled by volunteers who found that, although it had long been inhabited by raccoons, it was in very good shape. The 3000 components have since been re-erected at the Henry Ford Museum in Michigan as an example of what can, in theory, be done.

Life could be boring in poorly thought-out suburbs, with few facilities available unless the housewife had access to a car. Pregnancy was called "The Levittown look". One way to fill the void was to buy lots of consumer goods. Many of these were powered by electricity, and it was its early introduction to numerous American homes which made the products and hence the consumer age possible. Every increase in sales gave opportunities for manufacturers to cut costs and so prices, and hence to sell still more. From the 1920s consumer goods began to be sold in vast numbers. For example, one way to keep in touch was by using the telephone. The percentage of American households with a telephone was 41% by 1929, far higher than in any other country. With the Great Depression this figure fell and was not exceeded until 1942, and then continued to grow to 90% by 1970. The appetite in America for new products, especially for anything that saved time and effort, was enormous, and American industry was only too willing to supply them—if you had the cash.

Bluetooth® is the name for the concerted effort to use telecommunications to enable the user to communicate with all these household appliances. Ericsson was the Swedish instigator of the technology and named it in 1994 (after a Scandinavian king). Nine companies in the telecommunications and computer industries are working together to develop ideas and to establish a standard. This will avoid the mess experienced over, for example, rival video recording systems. It is still more an idea than reality. One problem is physically moving things around. Something might be in the freezer and you want it defrosted and then cooked in the microwave. No number of messages sent from work will sort this out unless, say, the microwave acts temporarily as a freezer until told to start cooking. Hundreds of patent applications mention the concept, which will almost certainly have real uses within a decade.

The kitchen is a particular target for such technology, with those returning from work or school not always content with fast food. With the housewife traditionally exerting so much influence over the purse-strings, much effort has been made to sell the American Dream for that room. US 3926486, which dates from 1972 by General Electric, looks somewhat old-fashioned now for those who like sleek lines everywhere, and the appliances hidden away, but the look is familiar to many. The title was "Modular furnishings" and the idea was to make it easier to install prefabricated cabinets. The patent comments that the construction of the kitchen is done piecemeal by the constructor during the building work. Plumbing comes first, then the walls, then the cabinets, and finally the appliances are installed. The plumbers and electricians then return to connect everything up. One apartment project was estimated to have taken, for the kitchen cabinets and appliances in one kitchen alone, 26 man hours over 3 months. Instead of building cabinets from scratch the patent proposed the use of eight varieties of prefabricated cabinets.

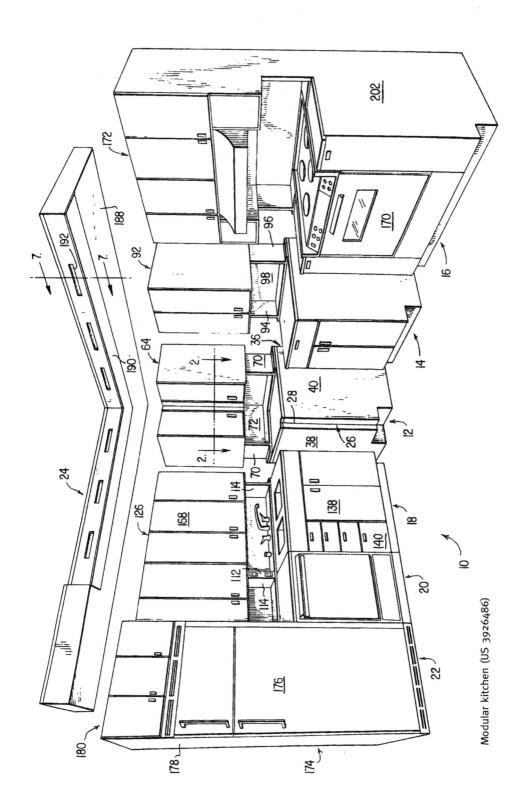

Modular kitchen (US 3926486)

These were meant to deal with the problems of the fiddly corner cabinets and to ensure a nice even line along the cabinets' edges. The basic idea of the patent was to install one of three kinds of corner units first whose front portions would then interconnect with the cabinets on either side. Each module would be in multiples of 1 foot (and one hopes that the appliances would fit accordingly). The patent stated that it was "highly desirable" that the electrical wiring was installed before the cabinets. Another suggestion was sliding doors above the sink unit. At (24) at the top of the drawing was a molding feature at the join between wall and ceiling, which provided lighting. This resulted in a perfect place for many a *Kaffeeklatsch* session.

The blender is thought to have originated with Stephen Poplawski of Racine, WI, for the Gilchrist Company in 1921 with US 1475197. This was for mixing drinks but it was the first to have a spinning blade at the bottom. The first to look and behave like a modern blender was by Frederick Osius of Miami Beach in 1937 with US 2109501. His "Disintegrating mixer for producing fluent substances" was named the Waring® blender in honor of bandleader Fred Waring. He had financed its development after being approached by Osius, who had talked his way into Waring's theater dressing room with an earlier (unsuccessful) model. Waring was an obvious contact as he was fascinated by gadgets, and he publicized the invention on the radio with a singing group called the Waring Blenders. Soon the blender was being purchased by numerous restaurants and bars. Unlike the Poplawski version it could cope with vegetables and fruit as well as liquids. The same model is still popular today.

The toaster took time to evolve into a useful device. Frank Shailor of Detroit in 1909 with his US 950058 for General Electric is credited with being the first toaster with the familiar wire coils for heating the bread, but it was primitive. You had to retrieve the toast when you thought it was ready. Charles Strite of Minneapolis with his US 1387670 in 1919 heated the bread for a set time before the toaster turned itself off, and this was improved by his US 1394450 which "popped up" the toast (using a timer and a spring). It sold for $13.50. For many years there was competition between "side loaders" and "top loaders" before the latter prevailed. Much innovation has occurred since. For example, modern toasters can often cook several slices at once, allow for preferred levels of browning, and for thicker items like beigels—not to mention crumb trays.

In order to get rid of the waste generated by cooking many take the garbage grinder for granted. The traditional model contains a circular disk rotating about a vertical axis in a cylindrical grinding chamber, often with hammers to break up the bones. Joe Shaver of Omaha with his US 1732775 in 1927, seems to have invented the first such device, but the first successful type was by John Hammes of Racine, WI, in 1933 with US 2012680. Hammes said in the patent that the garbage could be "delivered directly to the drain and sewer pipe", thus "eliminating the necessity of keeping the garbage in cans for collection with the contingent

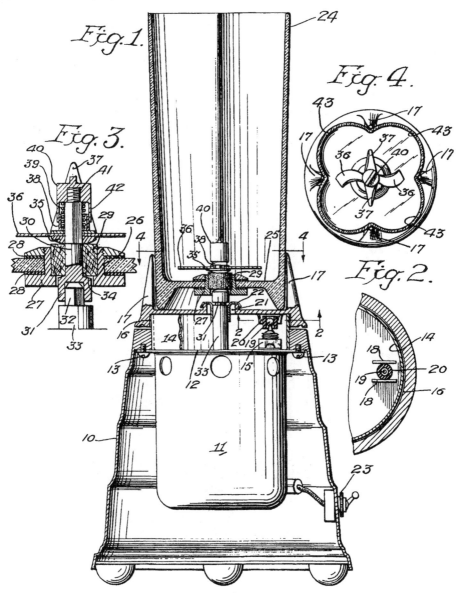

Inventor:
Frederick J. Osius
by his Attorneys
Howson & Howson

Waring® blender (US 2109501)

inconveniences". He suggested that the casing should be of aluminum, or some other non-corroding material. The workings are complicated and included a U-shaped grinder (27) with arms (28) and teeth (31) which cooperated with another set of teeth (32). The results went through a drain plate and out the drain (11) with a motor being housed at (24) at the bottom. Hammes, an architect, wanted to make the disposing of garbage easier for his wife. The first unit in mass production was General Electric's Disposall® from 1935.

Safety is very important, especially when small children are in the house. The first battery-fitted domestic smoke alarm is thought to be US 3460124 by the Interstate Engineering Corporation of Anaheim, CA, which dates from 1966. Hygiene or at the very least odor is also a concern if there are pets in the house. Kitty Litter® was a great solution to an unpleasant problem. Ed Lowe was born in St. Paul, MN, and his family later moved near Cassopolis, MI, where the family

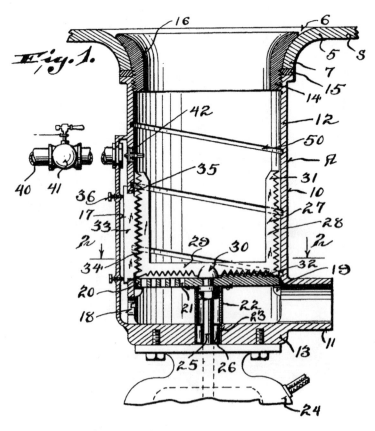

Garbage grinder (US 2012680)

business was selling industrial absorbents. This included sawdust and an absorbent clay named Fuller's Earth. One day in 1947 Lowe was approached by his neighbor Kay Draper who was tired of using ashes in her cat's box and hence dealing with sooty paw prints all over her floor. She asked for some sand, and Ed suggested Fuller's Earth instead. She had previously been using sand from a pile in her yard, but it would freeze in cold weather.

Soon the neighbor would use nothing else, as the clay was much more absorbent than sand and did not leave tracks all over the house. Lowe thought that other cat owners would like his new cat box filler, too, so he filled 10 brown bags with clay, wrote the name "Kitty Litter" on them, and called on the local pet store. With sand available for next-to-nothing, the shop owner doubted whether anyone would pay 65 cents for a 5 lb bag. "So give it away", Lowe told him. Soon customers were asking if they could pay to get more. Lowe began crossing the country in his 1943 Chevy Coupe visiting pet shops, or going to cat shows, where in return for cleaning hundreds of cat boxes he was allowed a stall where he could display his new product.

Although sales were good Lowe realized that his product was just clay, readily available to any manufacturer, and only the trademark was protecting him from competition. The original idea could not be patented but he obtained patents for additional ingredients to stem the growth of odor-causing bacteria or to reduce the dust in the cat box.

When he sold his company in 1990, Edward Lowe Industries was selling more than $210 million of cat box filler every year.

Finally, the kitchen may be used for ironing. Henry Seely of New York City patented the first electric iron in 1882 (US 2590054), which used any "suitable source of electric energy".

(A) and (B) were two separate parts and (C), with its electrical resistors (B), was in between. A "multiple-arc system of electric lighting" was suggested as the power source with the interior terminals (E) of a lamp socket being attached not to a lamp but to a plug (F). This was at a time when electrical supply was not supplied domestically, with Edison's first system for providing power to lower Manhattan only beginning in September 1882, several months after the patent was applied for. Gas lighting was available but of course no power points. Therefore Seely had to design an entire system, which included a way of avoiding short-circuiting the apparatus. The idea of the wedge shape was not new, as the advantage of being able to iron fiddly bits had already been appreciated with gas powered irons and those pre-heated in fires. Steam irons arrived in 1926, but were initially huge and were used only in laundries.

Turning to the living room, foreigners often joke that American living rooms are typically laid out so that everything faces the TV. Castro Convertibles Corporation of New Hyde Park, NY, filed in 1969 for US 3608101 which gets even closer to this aspect of the Dream, particularly in smaller rooms.

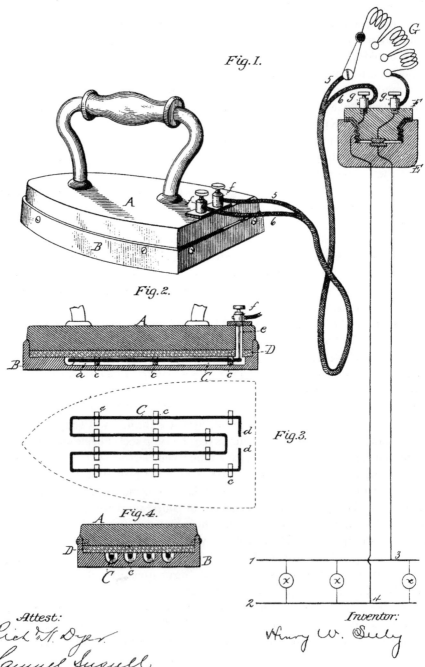

(No Model.)

H. W. SEELY.
ELECTRIC FLAT IRON.

No. 259,054.   Patented June 6, 1882.

Fig.1.

Fig.2.

Fig.3.

Fig.4.

Attest:

Rich͂ N. Dyer

Samuel Sunell

Inventor:

Henry W. Seely

Electric iron (US 259054)

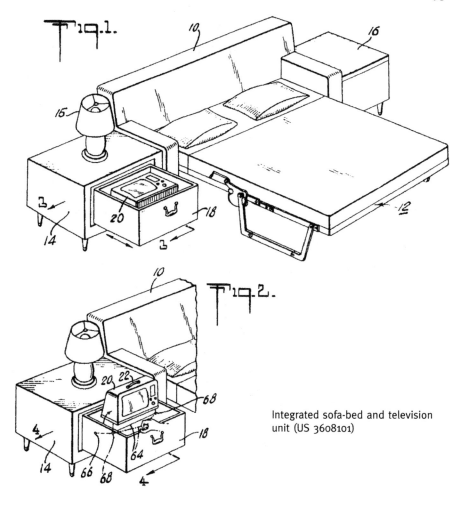

Integrated sofa-bed and television
unit (US 3608101)

It is for a sofa bed with a TV set hidden inside a furniture unit at one end, which makes it convenient for watching while lying on the sofa, though perhaps less so while lying in bed. A phonograph was meant to be in the unit at the other end. The patent comments that sofa beds with phonographs or bars in cabinets placed at the end were known, but here watching the television would be made more convenient. When not in use the TV lay on its back on a foam rubber mat within the drawer. After pulling out the drawer, the handle (22) at the top of the set was grasped and lifted so that the set locked into position. It could be then be swivelled to get the best view. To be fair this kind of living style is very uncommon, and if it were more usual electronics would be used for the hard work of putting the set in position.

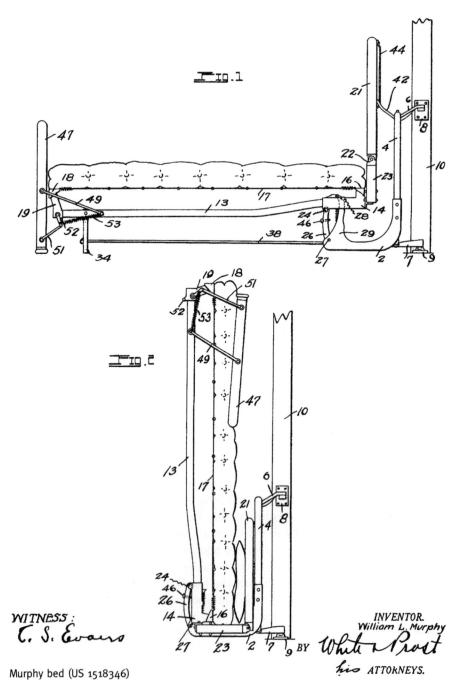

Fig.1

Fig.2

INVENTOR.
William L. Murphy

BY White & Prost

his ATTORNEYS.

WITNESS :
C. S. Evans

Murphy bed (US 1518346)

A more common way of saving space is by using a Murphy bed which springs up into the wall. William Murphy moved to San Francisco in about 1900 and lived with his wife in a one-room apartment. A bed took up most of the floor space. As they needed space to entertain he began experimenting with a folding bed. Despite claims on the internet this first version does not seem to have been patented, but a later version was. This was US 1518346 which dates from 1919. The bed is closed by pushing on bar (47) so that, while it swings over onto the end of the bed, a counter-balancing mechanism raises the bed up into the framework (2). Once folded the bed turns on pivots (7) and (9) so that it swings into a closet behind it, and the door on the other side then shuts on the closet.

Some 30,000 were sold annually in the 1920s and 1930s and this type is still in use today. Modern versions of the idea vary, but normally the bed is made to look like part of a cupboard or cabinet when not in use (sometimes doors have to be opened before the bed is lowered). Murphy did not invent the concept of a folding bed but his were the first to be popular. Earlier, Sarah Goode, who owned a Chicago furniture store, had invented the folding cabinet bed. This folded up against the wall into a cabinet and could then be used as a desk with compartments for stationery and writing supplies. Goode's patent, US 322177, filed in 1883, is thought to have been the first by an African American woman inventor.

The reclining chair is a great boon as it feels good to put your feet up. The Laz-Z-Boy® reclining chair is a well-known type and has been called the most comfortable chair ever made. The inventors were cousins Edward Knabusch and Edwin Shoemaker of Monroe, MI. Originally it was a wooden slat chair, but a customer had suggested that they made an upholstered version, and this was the kind which was filed in 1929 to become US 1789337. The user presses on the back so that the chair adopts either a reclining or an upright position. The trademark was also used from 1929, and was chosen after a contest (which gave welcome publicity). Laz-Z-Boy® won out over suggestions such as Sit-N-Snooze and Slack-Back. The cousins continued to patent into the 1970s with improvements such as extendable footrests.

Moving on to the bathroom, a great deal of attention has been paid by inventors to our comforts, partly because it can feel rather cold in there. The Japanese are prolific inventors in the field of toilet seat warmers, but Columbus Brandi of Tahoe Valley, CA, with his US 2717953 in 1953 predates them. Brandi remarked that his invention was removable, unlike "the other toilet seat warmers of which I have knowledge". There is clearly room for further research in this little-explored field. Brandi provided for a heating element encased in plastic which went underneath "an attractive protective covering" for the lid. Another device for warming things in the bathroom is the heated towel rail. The first such device involved circulating hot water through the rail and is attributed to Henry Keeling in about 1900. There is no end to inventors' ingenuity in the bathroom: even a toothpaste tube warmer has been patented. This was by Frank Williams of Baltimore with US

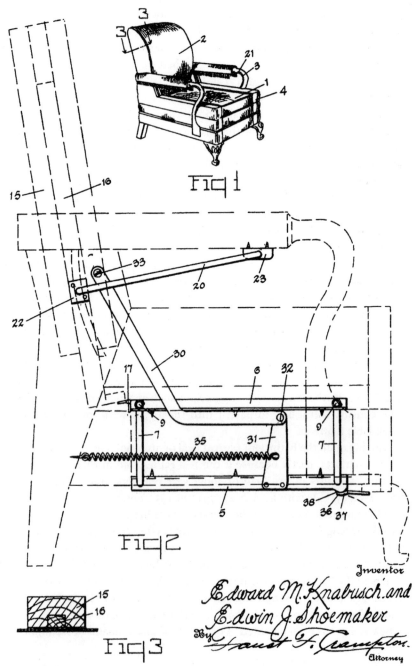

FIG 1

FIG 2

FIG 3

Inventor

*Edward M. Knabusch and
Edwin J. Shoemaker*

By *Faust F. Crampton.*

Attorney

Laz-Z-Boy® reclining chair (US 1789337)

6204485, filed in 1999. It is plugged into the mains and can be controlled by a timer. The idea is to help those who suffer from sensitive teeth.

Some baths are now adapted as whirlpool baths, often (but incorrectly) called by their owners after the Jacuzzi® whirlpool bath. This dates back to 1968 with Roy Jacuzzi of Little Rock, AK, for his Jacuzzi Corporation. It was the first self-contained, fully integrated whirlpool bath which incorporated jets into the sides. His US 3452370 was entitled "Hydromassage installation". An electric pump circulates the water through the pipes using an underwater suction fitting and several underwater jets. Each jet contains a venturi, a vented constriction near its opening that injects air into the water.

Beardless men are divided between those who wet shave and those who dry shave. Those who dry shave owe their product to Jacob Schick of Ottumwa, IA. He had been an army officer and after early retirement went looking for gold in Alaska and in British Columbia. He found it uncomfortable wet shaving when it was −40 °F weather, so he began to tinker with the idea of a shaver with a shaving head driven by a flexible cable and powered by an external motor. Manufacturers rejected the idea, as it was unwieldy, with the drive shaft separate from the head, each held in one hand. He served again in World War I where he became a Lieutenant Colonel. He was so impressed by repeating rifles that in 1921 he called a new product the Magazine Repeating Razor. Clips of razors were loaded into the handle so that a new razor could replace the old one without the user touching it. The product did well, but his real love was the idea of the electric razor. He eventually went on to invent the first electric dry shaver with oscillating blades with US 1757978, filed in 1928 from his Stamford, CT home. He had to sell the assets of his old company to finance production. The hairs but not the skin come into the gaps between the teeth on the blades. The patent said that each such gap was to be between 0.008 and 0.02 inch. This time he did not do well (the Great Depression cannot have helped), and rival products emerged in the 1930s from Europe as well as the USA to compete. Schick died in 1937, almost forgotten.

Toothpaste dates back to Britain in the late 18th century. It was very abrasive and included such ingredients as brick dust, china, earthenware and cuttlefish. Borax powder was later added to create a foaming effect. This may sound awful, but then few ever ask what is in the stuff today. Until 1940 soap was a common ingredient. It was not until 1956 that Proctor & Gamble introduced Crest® toothpaste, the first with fluoride. As for the basic idea of the familiar collapsible metal tube, this took a long time to arrive. John Rand, an American portrait painter who was living in London, England, at the time patented in 1841 with US 2252 his "metal rolls for paint". It did not occur to anyone to use the same idea for toothpaste until about 1892 when Washington Sheffield of New London, CT, began manufacturing toothpaste tubes (he called his product Dr. Sheffield's Crème Dentifrice). Previously pots had been used. The first competition to tubes was not until 1984, when the pump-dispenser was introduced to the USA by Henkel of

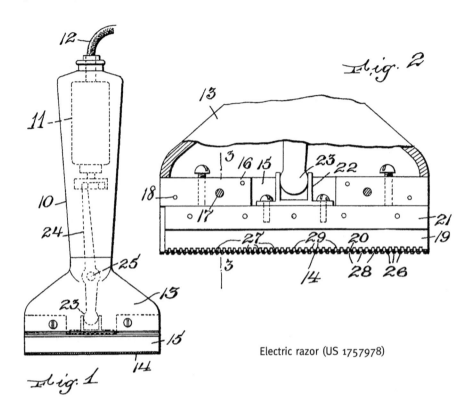

fig. 2

fig. 1

Electric razor (US 1757978)

Germany. Eighteen per cent of the US population are now thought to favor the pump variety. As for striped toothpaste, this innovation arrived with US 2789731 by Leonard Marraffino of Mount Vernon, NY, which was filed in 1955. There are individual "passages" in the tube for the different ingredients and they meet at a "striping port" where they are forced to merge into one stream.

The electric toothbrush is not just for ease of use by those who get tired of brushing, but also ensures efficient cleaning if used according to instructions. The idea is thought to date back to 1880 but the first successful product was US 3227158, which was filed in 1961. The inventor was John Mattingly on behalf of the Aquatec Corporation of Fort Collins, CO. It was a "Method and apparatus for oral hygiene". It became well known as the Waterpik®. The patent explains that while the idea of a liquid jet was known there were difficulties in "securing adequate and sustained liquid pressure and in providing simple and easily manipulated apparatus for nontechnical users"—presumably, most people. It goes on to say that extensive testing was carried out to determine the optimum factors. The best diameter of the jet was found to be between 0.032 and 0.038 inch, with water pressure at 90 lb/sq in which pulsed at 1,250 cycles per minute. The basic idea is that the motor-pump assembly (10) takes water from a tank behind it at (11) and through (12) discharges it to the nozzle (13). Mattingly was a hydraulic engineer and former professor who teamed up with Gerald Moyer, a dentist. They formed a start-up company, Aquatec, which is now Water Pik Technologies. The company has expanded from the original rented house and continues to make "oral irrigators". Flosser versions also exist, such as US 5572010, filed in 1993 by Dane Robinson, a Paradise Valley, AZ, "implant dentist".

Turning to the bedroom, you will spend a third of your life in bed and it is important that you feel comfortable there. The traditional coiled metal-spring mattress has a long history, dating back to an 1826 British patent by Samuel Pratt of London which he said was based on his own work and that of a mysterious unidentified "foreigner". There have been remarkably few changes since.

Nor has the problem familiar to the many people who have difficulty waking and sleeping gone away. Some cannot fall asleep while others cannot wake up, and some keep others awake by sawing lumber all night. There have been over 150 patents for methods of preventing snoring. These use a variety of techniques, as explained in US 6089232, filed in 1998 by Leonard Portnoy and Alex Farnoosh of Beverly Hills, CA. It states "A broad variety of intra-oral and dental appliances and devices

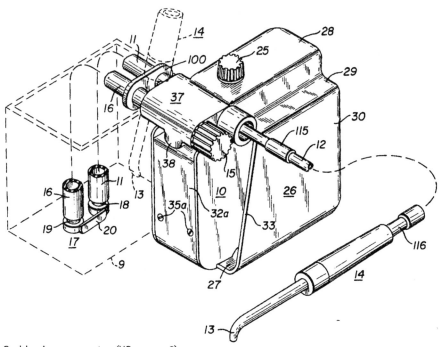

Oral hygiene apparatus (US 3227158)

are now available to treat a patient for snoring. The known oral devices for treating snoring and obstructive sleep are worn inside of the mouth and work by repositioning of the jaw, moving the mandible, lifting the soft palate or moving the tongue forward. . . . All of these known devices suffer from having to be worn inside of the mouth, they are uncomfortable to wear and they are expensive and must be fitted and made to order. These devices also often cause excessive salivation, dry mouth or tempomandibular joint (TMJ) discomfort." Maybe the cure was always going to be worse than the problem. The solution offered by the inventors was a peel-off, adhesive sheet of material which was placed over the face to cover the mouth but which still enabled breathing through a thin slit across the middle of the sheet.

Another idea was an Australian offering by Colin Sullivan, filed in 1981, which became US 4944310. It was for a mask which went over the face with two snorkels rising up like horns from the nose. The idea was to suppress snoring by using varying air pressure. Then there was US 6435905 with its "Multi-component snoring treatment", applied for in 1999 by Pi Medical of St Paul, MN, which consists of an implant in the soft palate. This is one of seven patents granted to the company on the subject.

A house needs maintenance, and power tools used to be huge, awkward things that were definitely not personal or portable. In 1913 Samuel Black and Alonzo Decker of Baltimore applied on behalf of their Black and Decker Manufacturing Company for US 1245860. It was the first (relatively) portable electric drill, weighing 24 lb, and costing a hefty $230. It was also the first which could be operated with one hand on the trigger while the other hand was free to say steady the drill. The patent states that previously it was necessary to move the hand to control the current which often meant that the drill bit was damaged if the drill sagged. "Our improved switch may be operated without in the least releasing the grip or support of either hand of the operator." There was no confusion about what switch to use: you simply operated the drill as if it were a gun. This was the beginning of the DIY revolution, which was also helped by such apparently simple items as the Parker-Kalon self-threading screw, the first screw with a pointed end, the patent for which was applied for as late as 1923 by Russian immigrant Heyman Rosenberg of New York City with US 1485202, and by the Phillips screw with its secure cruciform head, by Henry Phillips of Portland, OR, applied for in 1934 as US 2046343, besides many other more complicated devices.

Something in this area that badly needs solving is pouring paint from tins. It took until 1987 for a "spill bill" to be invented which could be fitted to a tin so that the paint does not go everywhere when poured. Why should paint tins not have simple spouts? The problem is that you may want to mix ingredients in the tin, or to stir it.

Few enjoy cleaning although for some it may be their main source of exercise. The ultimate in easy cleaning of the home, perhaps, has already been achieved by

1,245,860.

Patented Nov. 6, 1917.

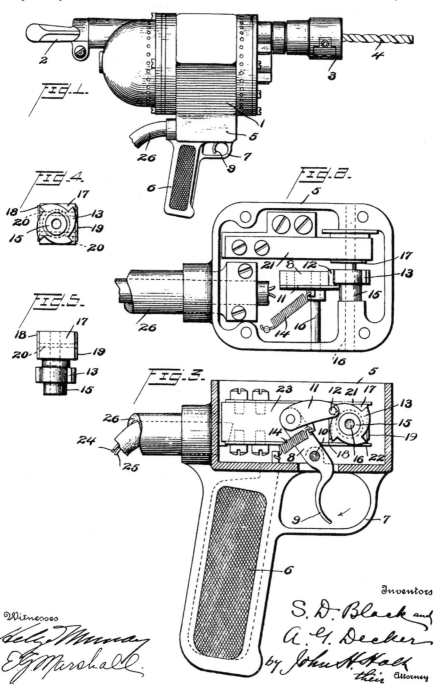

Witnesses

Inventors
S. D. Black and
A. G. Decker
by John H. Hall
their Attorney

Portable electric drill (US 1245860)

Frances Bateson of Newburg, OR, with her US 4428085, applied for in 1980. Her "self-cleaning building construction" is "characterized by ease of upkeep". She points out that although there are many labor-saving devices around the home, little has been done to automate cleaning. The invention is not confined to the house alone: dishes and clothes were also included. Bateson's question, "Why should people waste half their lives cleaning the house?", will be only too familiar.

The drawing shows a view from above of how a house would be organized according to her system. The small units (20) are spray apparatus suspended from the ceilings. "Baseboard device" units (22) at floor level deliver heated air to dry out rooms. The arrows in each room indicate the drainage slope, with 0.5 inch/10 feet being the suggested tilt. These drain towards either fireplaces or grills in the floor. Possible problems with her scheme include the householder having to get used to storing everything away before cleaning as money, check books and other papers left out would get ruined; protecting electrical apparatus against short-circuits; and the cost of providing so much plumbing and waterproofing.

Under the name of Frances Gabe, her invention has achieved much fame in inventors' circles. She went to college at 14 and earned a degree in 2 years. She learnt much about houses from her father, an architect, and used her "pen name" to avoid embarrassing her husband when for 35 years she successfully ran a housing repair business. Her own house in Newburg has been rebuilt to her principles

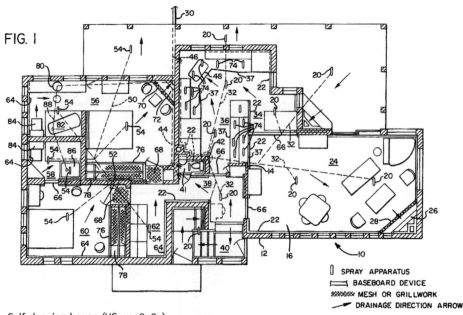

FIG. I

| | SPRAY APPARATUS
⊏⊐ BASEBOARD DEVICE
▨▨▨ MESH OR GRILLWORK
⤏ DRAINAGE DIRECTION ARROW

Self-cleaning house (US 4428085)

with valuable items kept under glass, and specially treated furniture. There are no carpets, which she considers dust traps, and all handbasins and sinks (and also the storage area for cutlery and crockery) are designed to be cleaned automatically. Clothes, too, are cleaned while still hanging in the cupboard.

Most people would find the orderliness of living in such a house too difficult to cope with. One idea in cleaning that has been much talked about but rarely if ever seen is that of a robotic vacuum cleaner. Again damage to the house is a real fear: the machine may run amok while coffee is being drunk in the next room. Again, how will valuable items be protected? Procter & Gamble are one company which have taken up the challenge. Their international patent application WO 01/37060, filed in 1999 by eight inventors from their Cincinnati base, is for a "Home cleaning robot".

The 75 page patent specification suggests that the robot can perform a useful function but can also be like a toy. After discussion of some interesting sounding inventions in the field the specification states, "There is an unfilled [need] for home cleaning robots that use low energy cleaning techniques and thus make chores easier for the user". The toy angle is apparently that children will enjoy watching the robot perform its work, particularly as it is suggested that the robot should be given the appearance of a mouse or a turtle. There is even the clever suggestion of using a "bunny rabbit with large ears that act to dust base boards". The actual cleaning is carried out with a non-woven electrostatic cloth. When the robot comes across an obstacle a sensor detects it and the robot reorients itself. Gradually a picture of the room is built up for reference. It is suggested that this "map" should initially be built up by a person leading the robot round the obvious parameters of the room. Most of the specification discusses how the robot will find its way around the room. It is designed to weigh less than 17 lb. Perhaps the toy aspect will not appeal so much to the father or mother who actually contemplates the purchase. An even lighter model, Roomba™, went on sale in September 2002. It weighs only 6 lb and was designed by iRobot of Somerville, MA. It works along similar lines but says "uh-oh" if it bumps into anything. No doubt children will deliberately maneuver the robot into things so that they can hear it speak. Perhaps the next step will be a robot cleaner that is able to clamber over tables and chairs, dusting as it goes.

Food, glorious food

M ANY inventions, not all patented or indeed patentable, have led to the provision of numerous new foods, drinks and methods of serving them. More than in any other topic, perhaps, what is now taken as Americana was invented or introduced from outside the country, or by "hyphenated" Americans. If so they have been zestfully adopted (like their immigrant originators) into the country. More ideas constantly emerge to stimulate jaded palates. Again and again these inventions have inched towards the American Dream by making life more convenient, faster, hygenic, and all round more comfortable—and tasty. Sadly most also have lots of calories, sugar or salt.

Not all the stories here can be verified, and many are doubtless "urban legends" to account for the origins of a name. If nothing else urban legends speak of the ingenuity somebody showed in coming up with such stories in the first place. If any food could be pointed to as essentially American, chewing gum would arguably be the winner. There is some controversy about its origins and who deserves the credit for its introduction. There were earlier gums that consisted of spruce or paraffin, but the first chewing gum made from chicle was manufactured by Thomas Adams, a photographer living in Staten Island, NY.

This was a result of the influence of a somewhat improbable neighbor— General Santa Anna, the victor at the Alamo in 1836. As he frequently made Mexico too hot to hold him, Santa Anna went several times into voluntary or involuntary exile and this was the 1866–67 period when, as a septuagenarian, he was staying in the USA (although his inability to speak English may have caused problems). He arrived with a ton of Mexican chicle, which he hoped to sell as a rubber substitute. Adams' initial interest in that idea changed when he noticed that his son had taken up chewing the stuff in imitation of Santa Anna. These were solid, unflavored (and unsugared) balls. Adams began to make and sell the product, with some success, his slogan being "Adams' New York Gum No. 1— Snapping and Stretching". From 1871 he pioneered the first addition of a flavoring, licorice, to a product which he called Black Jack. Adams is credited on several internet sites with patenting a chewing gum making machine in 1871, but in fact his US 111798 is simply for washing the "chickly" in two hot baths. In this way he claimed that "when softened by the warmth of the mouth [it] becomes very tenacious and ductile".

John Colgan, a druggist from Louisville, KY, is generally credited with improving the flavor of chewing gum. In 1880, he added the flavor to the sugar before the sugar was added to the chicle. This made the flavor last longer as the gum was chewed. Another early innovator was William Wrigley, Jr, a Philadelphian who in

1893 introduced Wrigley's Spearmint® gum. He was the first to add latex to make it stretch. The family business had tried offering baking powder as an incentive to retailers to buy their soap. The baking powder proved to be more popular than the soap so they began concentrating on that, offering this time packs of chewing gum as an incentive. That in turn proved more popular than the baking powder. Chicle is mostly replaced nowadays in gum by petroleum products and latex.

The credit for bubble gum, as opposed to chewing gum, is usually given to Frank Fleer's company. His 1906 Blibber-Blubber product had poor sales as it was awkward to chew, and tended to explode in the user's face before achieving a satisfactory size. He was another Philadelphian who went on to register over a dozen efforts at trademarks for chewing gum, all of which flopped. These included Blimp, Fruit Hearts, Vi-lets, Whiz Bang and (preceded by the word Fleers) Gumps, Coolmints and Treasure Hunt. Another predictable failure is illustrated.

Finally Fleer was successful with what was the first true bubble gum, and at the same time his first popular trademark: Dubble Bubble® gum. Walter Diemer, of his Frank H. Fleer Corporation, created the new (unpatented) formulation. The first batch was pink as that was the only color available when they wanted something eye-catching, and has remained so since. A legend claims that somebody's drawers fell into the vat and that the workmen were too scared to explain the reason for the color.

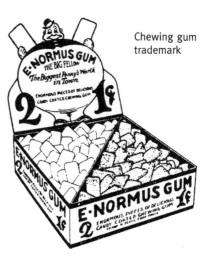

Chewing gum trademark

Cooling drinks and food have been important in a country with hot and humid summers, and hence have contributed to that part of the American Dream which involves being comfortable. Ice cream is an obvious example of such a food, and shows how innovation steadily improved what was available to the public. Ice cream was originally a great luxury due to the problem of keeping it cold enough before cheap refrigeration was available. Dolly Madison is generally credited with introducing ice cream at her husband's second inaugural ball in 1812. Actually, ice cream had already been enjoyed half a century earlier. In 1744, a dessert called strawberry ice cream was served at the Governor's Mansion in Annapolis, MD. We also know that George Washington bought a "cream machine for ice" in 1784 to use at Mount Vernon, and that Thomas Jefferson brought back a recipe from France in 1789.

Ice cream's emergence as an every-day product dates from the work of Nancy Johnson. The first (hand-cranked) freezer was patented by Nancy Johnson of Philadelphia with US 3254 in 1843. Previously, she explained, a vessel containing a mixture was rotated by hand within an outer casing containing salt and ice. Her much more economical method was to have a vertical shaft within the inner vessel which projected out beyond the lid with a handle attached. Two perforated "wings" were attached to the shaft. The outer vessel was as before and could, she noted, be shrouded with blankets to reduce the loss of ice. By turning the handle, the entire mixture came into contact with the outer wall and became frozen. She suggested using a glass cylinder while the wings could be of hard wood or ivory, or, if the mixture were not acidic (such as citrus fruit) of tinned iron.

Much of the selling (and inventing) in the ice cream field was carried out by Italian immigrants and their descendants. Italian immigrant Italo Marcioni of New York City invented a way of making cones from a mold in 1896 although he did not apply for his US 746971 until 1903. He explained in the patent that "ice cream cups" were formerly made by two adjacent blocks which both held molds. The blocks were separated and the "dry, crisp contents" removed. Basically, his invention consisted of making the molds hinged and in parts so that it was much simpler to open them. The side blocks could be removed while pins (g) were used to move the blocks around, and a latch (k) locked the molds in place. These early cones were made of pastry, and the stiffer waffle-like cones used today are said to have originated from an incident at the St Louis World's Fair in 1904. Ernest Hamwi, a Syrian immigrant, was selling a thin Persian waffle called a Zalabia. A nearby ice cream seller ran out of dishes, so Hamwi offered to wrap the waffle round the ice cream. The idea was immediately popular—you no longer had to hang around so that you could return the dish—and the treat was dubbed the World Fair's Cornucopia.

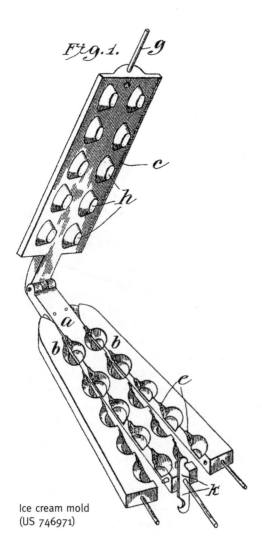

Ice cream mold
(US 746971)

Chocolate-covered bars were introduced by a Danish immigrant, Christian Nelson of Onawa, IA. He was a teacher who ran a candy store in his spare time. An eight-year-old came in one day and could not decide between spending his money on a chocolate bar and an ice cream. Nelson decided to combine the two. He filed in 1921 for what became US 1404539, the drawing for which predictably simply shows a dark bar. Half the rights were assigned to Russell Stover of Chicago, who was presumably a backer (Russell Stover Candies is still around as a company). Nelson wrote in the patent of his "frozen dainty" with an "encasement therefor with facilities for ready handling", which he noted further on in the specification was, of course, "edible". He began to call it the I-Scream bar after a song he wrote: "I-scream, you scream, we all scream for ice cream". On second thoughts, thinking that if the tune lost popularity so would his product, he renamed it Eskimo Pie®. When he died in 1999, aged 92, the product was still the leading brand in the field.

Harry Burt of Youngstown, OH, applied in 1922 for US 1470524 which resulted in another famous brand. The patent involved pressing sticks into a block of (unfrozen) ice cream and, after freezing, dividing it up into separate portions for sale. The patent makes it clear that hygiene was foremost in his mind. Trademark records show that he was using Good Humor® as a brand name for his "ice cream suckers" from December 1921—although winter was perhaps not the best time to launch such a product. The actual Good Humor Man vans began operating in 1923.

Another product originated with Frank Epperson, a lemonade-mix salesman from Oakland, CA. He had been demonstrating his product in New Jersey. One cold night he accidentally left a glass of lemonade with a spoon in it out on the window sill. In the morning Epperson pulled on the spoon and found that he was holding the first popsicle. He applied for a patent in 1924, US 1505592. He too stressed the convenience and hygiene of the product: it could be "conveniently consumed without contamination by contact with the hand". His idea was different and more cumbersome than Burt's. He suggested using small containers such as, of all things, *test tubes* which were to be filled with liquid syrup, with the sticks pressed down to the bottom to overcome buoyancy in the liquid. They were then quick frozen within a few minutes. Another explanation for the invention's origin exists. This says that he invented it when only 11 years old, in 1905. He had left his fruit-flavored punch out on the porch with a stir stick in it. His children are credited with suggesting the name. What is definite is that the Popsicle Corporation registered the word Popsicle® as a trademark for "lollypops", claiming its use from May 1923—before the patent was applied for.

From ice cream we can turn to candies. American consumption in 2000 was some 7.1 billion lb, about half being chocolate, with a retail price of $23.8 billion. The field was and is dominated by the great names of Mars and Hershey, two companies that originated as family-owned companies (as Mars is still). Together they account

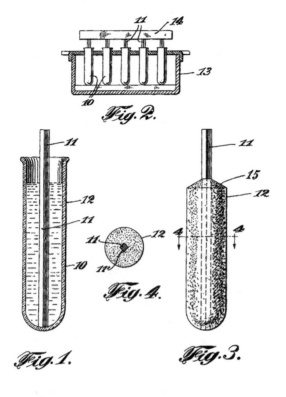

Fig. 2.

Fig. 1.

Fig. 4.

Fig. 3.

Popsicle® sticks (US 1505592)

for over half of American candy consumption, as more and more of the 6,000 companies that were in the field in 1945 get taken over. The histories of these two companies have, however, been quite different.

Milton Hershey was born in 1857 near Derry Church, PA, to Mennonite parents. He was only educated to the fourth grade, and was apprenticed to a candy maker. After many reverses he became from 1885 a successful maker of caramel candies. He became interested in the German chocolate-making equipment displayed at the Chicago International Exposition in 1893 and began producing chocolate covered caramel, starting his Hershey Chocolate Company, originally as a sideline. In 1900 he sold the caramel business for $1 million and began to concentrate on his new product, the Hershey Bar®. It was only in 1968 that Hershey Foods discontinued the nickel Hershey bar sold since 1894. The size had shrunk from 1.12 oz in 1947 to 0.75 oz, and Hershey announced that further size reductions were impractical. The new 10 cents Hershey bar weighed 1.5 oz.

In 1903 Hershey began to build a factory near his birthplace to take advantage of the plentiful supply of fresh milk for his chocolate. He built houses for his workers and a town soon grew up, with Derry Church being renamed Hershey in 1906. Milton Hershey refused to advertise his product, convinced that quality would speak for itself. Despite this sales rose to $20 million annually by 1921. Other well-known products by Hershey Foods include Chocolate Kisses®, first marketed in 1907, Mr Goodbar® (1925) and the Krackel® Bar (1939). Some of his later ideas were less successful, such as onion and beet flavored sherbets. Hershey and his wife were childless and gave so much to charity that when he died in 1945 his estate was only worth $20,000. The main beneficiary was the Milton Hershey School, an industrial school for orphans, which still controls a majority of the voting rights stock.

From 1950 the company offered tours of their factory at Hershey, with a million visiting annually. They could see the chocolate flowing like molten metal into vats, and bars being made with "Hershey" stamped into the chocolate. Jöel Brenner in her book *The emperors of chocolate* says, "In elementary school, it was thought customary to eat a Hershey Almond bar by biting the letter *H* first, and then the letter *Y,* and then the *E* and the *R,* leaving behind the *SHE.* No one knew why children nibbled the chocolate this way; they just did. Like playing hopscotch or kick ball or eating hotdogs, it was part of growing up in the USA."

It was only as Mars overtook them in sales that, from 1970, the company began to advertise for the first time. Hershey did have one spectacular coup in the struggle for dominance with Mars. They took a risk and paid for Reese's Pieces®, which is made of peanut butter and Hershey milk chocolate, to be featured in 1982 in a low-budget movie without any famous stars called *E.T.* It is the scene where 10-year-old Elliott succeeds in attracting E.T. in the forest by scattering the product on the ground. Sales went up 60% following this "product placement". Mars had originally been approached in the search for colorful candies, but had declined. The H.B. Reese Candy Company, the manufacturer of the original Reese's Milk Chocolate Peanut Butter Cups® on which Reese's Pieces® is based, had been taken over by Hershey in 1963. Just before then, in 1962, they had filed for US 3108679

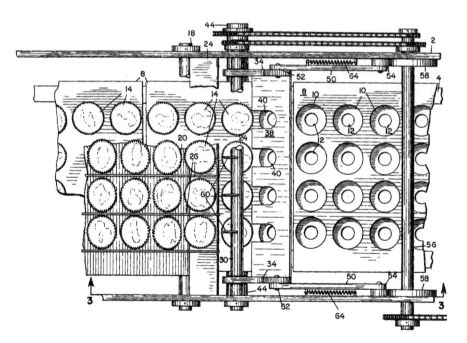

Making Peanut Butter Cups® (US 3108679)

by George Woody of Atlanta, GA. It shows the paper cups (14) being transferred from recesses in the trays (8) after cooling to a packaging area. Other drawings and a complicated text clarify for experts the exact way the conveyor belt system works for "successively fed articles".

Things were different at Mars. Frank Mars's childhood polio kept him at home a great deal where he learnt about making candy from his mother. He moved his candy making operation to Minneapolis in 1920, and soon devised both the Snickers® bar (without the outer layer of chocolate) and the Milky Way® bar. It was only in 1930 that Mars added the chocolate to both Snickers® and the Mars Almond® Bar. Named for a horse the family owned, Snickers® is the best selling candy bar in the USA. Frank Mars was pleased with how his business was going.

Forrest Mars, son of Frank, had much greater ambitions. "I told my father to stick his business up his ass", Forrest later recalled. "I wanted to conquer the whole goddamned world." He went to work for the Swiss firms Tobler and Nestlé, and learned some of their secrets. In 1932 he opened a small factory in Slough, England where he produced Mars® bars and developed the distinctive management style he retained all his life. Mars demanded complete loyalty and devotion from his employees. There were no private offices, executives had to punch in on a time clock like everyone else, and they could only travel economy. In return they were rewarded with profit-sharing and high salaries. Mars was famous for his temper, flinging boxes of candy around the factory floor if he noticed a single defective bar, or incorrect lettering on packets of M & Ms®. He added high quality pet food to the company's products, an area the company still dominates. All foods had to be sampled by the managers—including the pet food. "If we don't taste it ourselves, how do we know we're offering the best product we can?", asked one cowed manager. Having succeeded in England, Mars returned to America, set up shop in Chicago, and in time took over his father's flagging business from his father's widow, Frank having died in 1933.

Forrest Mars gave just two interviews between 1930 and his death in 1999. Because the company is privately owned, virtually nothing is known of its workings and finances. There have been, however, over 350 patents by the company which reveal many of their ideas. Visitors were *not* allowed into the Mars factories, but if they were they might have seen operations such as the one described and illustrated in 23 pages in US 2612851 by Robert Morrison of Oak Park, IL, which was filed by Mars in 1947. The title was "Candy making equipment". What is clearly a Snickers® bar at the top shows the candy bar that would be used in the particular format displayed (although it is not mentioned in the patent). The patent explains how nougat "dough" was mixed and whipped before the hot mixture was dumped into a hopper so that it came into contact with chilled rollers. Air was then whipped in to "preserve the light consistency". Caramel, nuts, and then more caramel was fed in to form layers. It was emphasized that keeping the layers to the correct height was essential (otherwise the wrapping would not fit). The

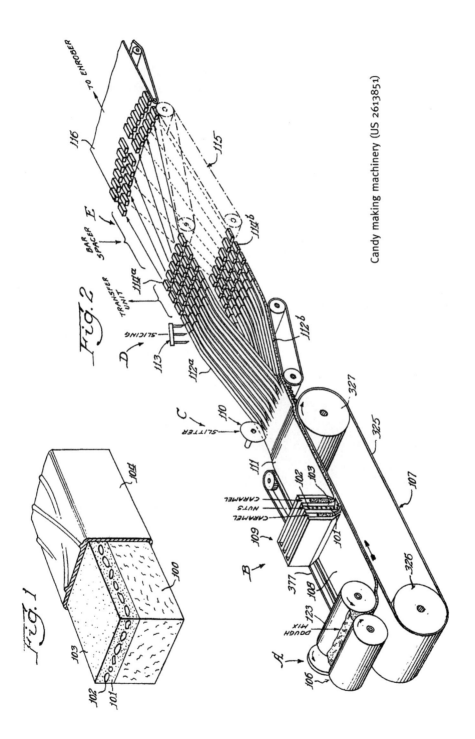

Fig. 1

Fig. 2

Candy making machinery (US 2613851)

mixture was then divided up and moved on out of sight to the "enrober" where the chocolate coating was added.

One current competitor to these great names is the Curtiss Candy Company of Chicago. Their most famous product is the Baby Ruth® bar, the origins of which are shrouded in mystery. This peanut and caramel bar wrapped in chocolate was introduced for a nickel in 1921 when the competition sold for a dime. The testimony when they applied for the trademark was that the mark was first used in November 1921. It is said that numerous children sent Babe Ruth the wrappers, asking for his autograph, and that Ruth demanded royalties for this use of his name, but was turned down in the courts. What is definite is that in June 1926 Ruth applied for a trademark for "Ruth's Home Run" candy, claiming its use since April 1926. The illustration is taken from the publication of the "advertisement" for opposition purposes by the Patent Office. This enabled anyone who objected to oppose the registration by the George H. Ruth Candy Company. The mark was refused, and an appeal by Ruth was lost in May 1931. In that court case Curtiss claimed that they had used their trademark since 1919 (not 1921 as had been claimed when they applied for it) and that they had spent millions of dollars in advertising, sometimes achieving sales of $1 million a month.

Ruth's Home Run trademark application

Curtiss have since claimed that the bar was named for Ruth Cleveland, daughter of the 22nd (and 24th) President, following a visit by her to their factory. As she died in 1904, and the company only began operations in 1916, this does not sound likely. If nothing else, using a trademark so close to the nickname of the "Sultan of Swat" must have been a great advantage in the marketplace.

The development of oven-cooked or microwaveable fries has made them more popular in the home, where many did not want the bother (and possible danger) of cooking in hot oil. French fries are hardly a snack for dieters as even a small portion will contain over 220 calories. Sugar is often added if only to even out the sugar level in the original potatoes, as it varies within the season when the potato is harvested.

Microwaves also helped with another product, popcorn. Many packs of microwavable popcorn have patent numbers on the packaging to cover the construction of the bag. Typically a bag is placed with the flapped side downwards and as it is heated the bag swells as the corn "pops" to become popcorn. Users are advised to stop the microwave when the pops become infrequent as overheating spoils the product. An example is US 6077551 by Hunt-Wesson of Fulleron, CA, which was filed in 1998. Before heating the bag has concertina-like creases at each side which open up to allow expansion of the popcorn. It can be torn open so that it becomes a "serving vessel". The inventor, Cynthia Scrimager, was well aware of the invention's contribution to the American Dream. "Microwavable popcorn has become a very popular snack item, convenience being a major contributor to its popularity. Microwave popcorn can be stored in a ready-to-use, shelf-stable, leak-proof package, which also serves as the cooking container". It might be thought that there were a limited number of variations on such ideas, but the US Patent and Trademark Office has granted over 100 patents on the subject. Many are classified under 426/111, which is for food packaging "having telescoping feature to increase or decrease package dimension or having packaging structure cooperating with food expansion".

Another famous company in the fast food business is KFC, formerly Kentucky Fried Chicken. The recipe for the coating used for their chicken is a secret, although attempts have been made to analyse it. What is not a secret is the basic way of

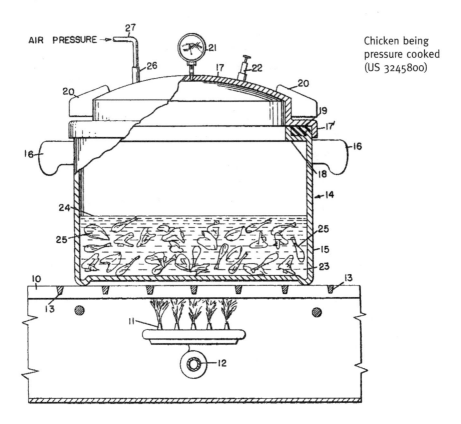

Chicken being
pressure cooked
(US 3245800)

cooking it. Harland Sanders of Shelbyville, KY, filed in 1956 for what emerged a decade later as US 3245800, a "Process of producing fried chicken under pressure". The patent explained that a consistent result would come from his method of cooking 2.5 lb chickens divided into 8 to 10 portions. The portions are dropped into fat at 350 to 400 °F. In 1 to 2 minutes the cold chicken lowers the temperature to 250 °F. This temperature is maintained for 8 minutes at a pressure of 15 lb/sq in. At that pressure, the boiling point rises to 250 °F rather than the sea-level 212 °F. The higher temperature means that cooking is faster (2 to 10 times faster, the patent claims) and the chicken does not dry out. Sanders was indeed the same famous "Colonel" Sanders who began a catering business in the dining room attached to his service station in 1930. By 1952 he was franchising his business, and in 1964 he sold out his interest in the Kentucky Fried Chicken Company for $2 million.

Liquid refreshment normally accompanies a hot snack. Coca Cola® is the only product to have two registered trademark names, the other being of course Coke®. Coca Cola® was registered in 1893 while Coke® was only registered in 1945, when it was claimed that it had been used since 1941, though it had clearly been used much earlier. There have been many court cases to defend these brand names such as one that ended up in the Supreme Court in 1920, Coca-Cola Co. v Koke Co. of America. The defendants were using Koke as a rival brand name. The Court said, "It is found that defendant's mixture is made and sold in imitation of the plaintiff's and that the word 'Koke' was chosen for the purpose of reaping the benefit of the advertising done by the plaintiff and of selling the imitation as and for the plaintiff's goods. The only obstacle found by the Circuit Court of Appeals in the way of continuing the injunction granted below was its opinion that the trade-mark in itself and the advertisements accompanying it made such fraudulent representations to the public that the plaintiff had lost its claim to any help from the Court."

In other words, was the company's spelling as Coke implying that cocaine was in the product? Before the case was brought the company had advertised that cocaine was not in the product (although it had been until 1903). "It appears to us that it would be going too far to deny the plaintiff relief against a palpable fraud because possibly here and there an ignorant person might call for the drink with the hope for incipient cocaine intoxication." So Coca-Cola won the case.

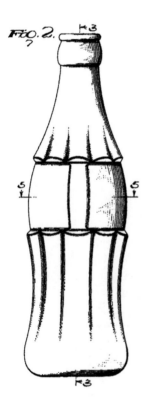

Coca-Cola® bottle
(US D105529)

Famously, the formulation for Coca Cola® was never patented and is therefore still a trade secret. The familiar waisted bottle originates in a company lawyer's suggestion in 1915 that the company adopt "a bottle which a person will recognize as a Coca-Cola bottle even when he feels it in the dark". Previous bottles had straight sides. Accordingly, in that year US D48160 was registered by the Root Glass Company of Terre Haute, IN. It was modelled on the kola bean and, while fluted, was a huge, bulging thing. It continued to be manufactured in several colors until 1928. A more familiar variation, US D63657, was applied for by Chapman Root himself in 1922. This went into production for a decade but lacked the deeper fluting and the slightly bulging middle of US D105529, which was applied for in 1937. The designer was Eugene Kelly of Toronto (but recorded as an American citizen). Besides being easy to grip, the bottle virtually sells itself by its high recognition rate, which is why it was eventually reintroduced after production ceased in 1951. Ingeniously, the company then secured in 1960 a trademark for a (modified) appearance of the bottle, complete with "Coca-Cola"

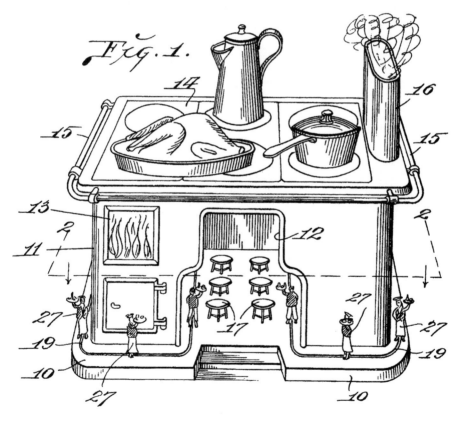

Conceptual restaurant (US 1795791)

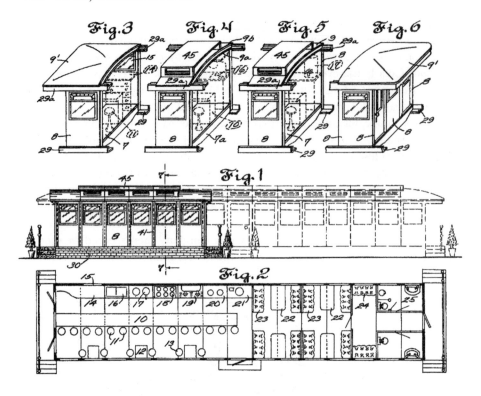

Railway diner (US 2089058)

in the well-known script across the middle, stating that "the trademark consists of the distinctively shaped contour, or confirmation, and design of the bottle as shown" and claiming use since 1916—which meant that such bottles should have been around then. These attempts to protect the company's usage of such bottles have clearly been very profitable even if cans are now cheaper to make.

The design of restaurants can also be a factor in inducing customers to buy. Only 60% of food expenditure is on food prepared in the home and an attractive venue is an incentive. Few, though, went to the extent of the delightful US 1795791 by Frieda Neuhaus of Los Angeles, filed in 1929, which was in the shape of a stove. The little figures on grooves were not the staff but rather stylised figures which were there for "attracting the attention of passersby and affording interest and amusement to patrons of the café or restaurant". The window (13) was meant to have air blown upwards to simulate flames playing over colored fabrics. Actual smoke from the kitchen came out of the chimney (16). This very expensive conceptual restaurant is not believed to have ever been built.

The "railway diners" concept was very popular in the 1920s and 1930s and some still survive as working businesses. Their sleek, modern lines attracted customers while for the owners their portability meant that they could be moved as business dictated as long as connections to utilities were available. One of many variations is attractively shown by Bertron Harley of Saco, ME, with US 2089058, dating from 1934. The patent enables the walls and roof to be interchangeable and adaptable so that, for example, windows and exhaust fans could be easily moved around,

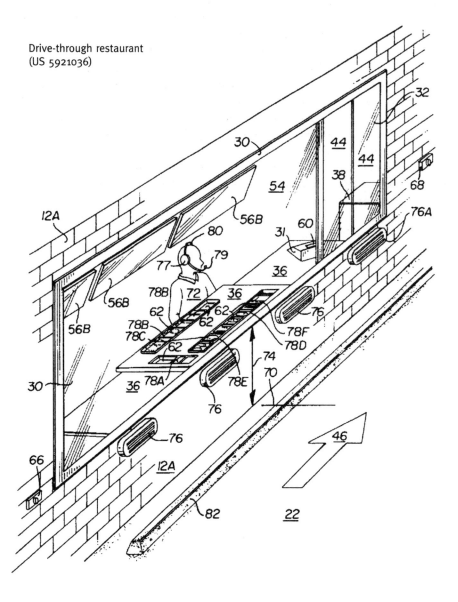

Drive-through restaurant
(US 5921036)

or the premises could be gradually expanded. Another factor, Harley said, were restrictions imposed by some states on transporting complete portable buildings by road. The drawings show "exploded" views, from the side and from above.

So far we have seen many examples of fast food. The granddaddy of a slick, professional operation serving fast food meals (rather than snacks) has to be McDonalds, although the White Castle® chain did precede them. Not just because of its huge share of the market, but because it is where the idea of a reproducible and consistent formula for fast food really began. Despite the early emphasis by McDonald's on drive-ins, drive-thrus are now much more common where land is expensive. A refinement designed for the drive-thru is shown by Michael Murphy of Montevideo, MN, in 1996 with his US 5921036. The idea was that car occupants could see what was being prepared before leaving the premises and hence saving on parking space. Murphy emphasized the need to have the car raised on the driveway so that the preparation work of "sub" sandwiches could easily be seen. This would mean fewer queries and waste of time while the bag was being inspected and also supplementary orders, hopefully, while waiting for the food. Hence the containers (62) and (78) were placed on the counter between the food preparation counters. Sensors would raise lighting when cars approached and reduce it as they left. The menus were above at (56B), while intercoms for ease of conversation were at (76).

After all this food we should be grateful that there are over 300 American patents for toothpicks. The first listed is by George Clark, Jr of Boston in 1875 with his US 174619. It is for a perfumed toothpick.

Here's health

A VITAL part of the American Dream is about living longer in order to enjoy life and its fruits. It is also about feeling fit and looking youthful, perhaps with enhancements to what Mother Nature has provided. Educational efforts have always been a strong theme in health. These include a number of patented board games with medical themes which make claims to be educational. One of these is US 5228860, entitled "AIDS: the epidemic board game". It was filed by Steven Hale of Orlando, FL, in 1992. Other board games cover such ideas as dentistry (US 4199145), sex education (US 4273337), weight loss (US 5704611), diabetes (US 6279908) and senior health care (application US 2002/0101032). Most unusual of all is perhaps US 5295694, the "Laparoscopic surgery simulating game". This refers to the procedure of making a small hole in the body and inserting an instrument to examine organs. The patent discusses in detail the Milton Bradley game called Operation®, where Cavity Sam is operated on by electric probes. The technology in that game was based on a 1965 filing by Marvin Glass Associates of Chicago, US 3333846, which is delightfully illustrated (but in a Western theme). By patenting, of course, all these inventors were showing they were not averse to making money out of their games, and many other games have been improvised by health promotion bodies but were not patented (and hence perhaps not recorded).

The idea of the "Vibratory weight reducer" round the body dates back at least to 1947 with William Rohlffs of Lancaster, PA, with his US 2549933. The "removal of surplus flesh" is the blunt intention of this invention. It worked by attaching the belt to a washing machine—but at what point of the cycle? The general concept of vibrating belts was popular in the 1950s in gyms. Now the electric belt has given new life to the idea as sold in countless catalogs. An emergency procedure is the adjustable gastric band. Professor Dag Hallberg of Sweden filed for WO 86/04998 in 1985 which was not subsequently given rights in the USA. The band is implanted in the body. Afterwards, a subcutaneous port is used to insert fluid into the band and gradually to increase the tightness. The first band was implanted in 1985 and since then more than 30,000 bands have been implanted worldwide. Previous work had not involved adjustable bands and further surgery was often needed. It has been compared to a strait-jacket for the patient. Less drastically, many weight loss formulas proudly claim patent rights and say that they have been "approved" by the Patent and Trademark Office, or "awarded" a patent. The gullible may think that this implies a recommendation of its value. It does not, as it just means that they think it is new. Patents do not have to work. More gentle in its approach is an invention by Milton Bruckner of

Los Angeles. In 1995 he filed for US 5522401, a "Stomach muscle/posture monitoring belt", which makes a clicking noise when the stomach muscles expand beyond a preset point.

Besides an addiction to fatty foods, other legal substances that are bad for you include recreational drugs such as caffeine, alcohol and tobacco. For a long time smoking was thought to be a healthy activity, and was even recommended by doctors. A major contribution to damaging health was the famous Zippo® lighter, the first windproof lighter. It was applied for in 1934 by the Bradford, PA, company as US 2032695 which was the model used by so many GIs. Another contribution to smoking was US 2163828, dating from 1937, the British invention of the "flip-top" cigarette packet. Above all, massive advertising, and the constant trickle upwards of smoke on many a movie screen, made the idea of smoking romantic and sophisticated, with a distinct erotic undertone.

It took some time for patents that discouraged smoking to make an entrance. These mostly rely on gradual withdrawal rather than going "cold turkey". Some of the earliest of these were "Time regulated cigarette dispensers" such as US 2953280 by Ernest Scarboro of Atlanta, which dates from 1958. A watch was included which would trip a lever at the set times. Timings could be from 30 minutes and is shown in the drawing as every 2 hours. A cigarette would then be removed from an "ejection tube" (300) at the bottom of the drawing.

However, Scarboro omitted to allow for easily reloading his device. Only one cigarette could be loaded at a time in the insertion tube (100) at the top to join the load of 30 (either king-size or regular, he states). In 1971 US 3722742 by Keith Wertz of Burbank, CA, objected strongly to this. He also felt that the timing mechanism was expensive and apt to fail. Another problem was that the timer only unlocked the dispenser for a short period of time. Having to wait for the next timed opening "could be very disturbing to the user". Needless to say, his patent solved the problem. The whole case could be filled at one go and the timer was not actuated again until a cigarette was manually ejected by using a "manual slider", which wound up the timer for the next cycle. There have been at least a score of patents on the subject.

Using a different approach, US 3402724 was filed in 1965 by two California inventors and is a device which fits onto the cigarette and gradually reduces the amount of smoke entering lungs as air takes over. On a more humorous note, US 3655325, from 1970, was a "pseudo-cigarette package that produces simulated coughing sounds when the package is picked up by a potential user". A battery-driven disk played a recording through a miniature loudspeaker. A self-extinguishing cigarette, US 4044778, was meant to go out if it went onto a surface. The patent examiner when listing similar inventions to it cited a *Mad Magazine* compilation from 1966 as "prior art". And then there were US 2681656, an ashtray where an elephant hoses down the fire, and US 4408620, where miniature firemen do the same job in a very elaborate invention.

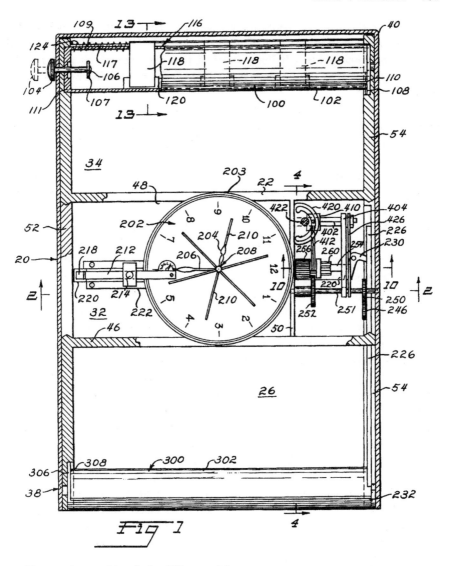

Fig 1

Discouraging smoking device (US 2953280)

More than 3 million people have died in accidents on America's roads, many more than have died in all of its wars. Many of these die due to drunk driving. At first the only way to determine the level of alcohol in a driver's bloodstream was through blood or urine tests, which were time-consuming and expensive. They could prove that someone was drunk at the time of an accident but did not help getting drunk drivers off the road before they had an accident. The first usable machine to check for sobriety was the Drunkometer, which was first used from

1938 (on New Year's Eve, appropriately enough). Rolla Harger, a biochemistry professor at Indiana University, had applied in 1936 for US 2062785. It contained the idea of blowing into a balloon attached to a machine which analyzed the alcohol in the breath. A color then showed which the police officer had to try to analyze by sight.

The famous Breathalyzer® machine dates back to Robert Borkenstein, also of Indianapolis, a retired Captain in the Indiana State Police, where he had been Director of the Police Laboratory. In 1954 he filed for US 2824789, "Apparatus for analyzing a gas".

The driver breathed into a rubber hose (16 and 17) and the air went into cylinder (13), with a handle (19) controlling its valves. Light (21) went on when the cylinder was full and turning the handle again released the air into a solution which analyzed it and recorded the results on a meter (33) after the color change was analyzed. Borkenstein was later a professor of forensic studies for decades at Indiana University, with a class on alcohol and highway safety attended by many police officers which is now simply called "the Borkenstein course". He said in 1995, "If we can make life better simply by controlling alcohol, that's a very small price to pay. My whole life's work has been spent trying to make life better for people." Among his other gadgets was a coin-operated analysis machine in 1970, which was meant to be installed in bars. You blew into a straw and for a quarter the machine would analyze your breath. Depending on your score, a message would flash, "Be a safe driver", "Be a good walker" or "You're a passenger". Just possibly some might have used it as a joke, and have drunk to enhance their score.

Borkenstein's technology was superseded in the mid-1980s as new devices use infrared light to test for alcohol in breath samples. A narrow band of infrared light, in the frequency that is absorbed by alcohol, is passed through one side of a breath sample. How much of this light makes it to the other side of the sample without being absorbed tells you very precisely the concentration of alcohol in the sample. The technique is expensive and the mechanical elements can mean problems, so the search is on for newer technology, perhaps using electrochemical (fuel) cells.

Polio has been a terrible scourge in the past. Philip Drinker and Louis Shaw were engineers at Harvard Medical School who wanted a machine that would sustain breathing for victims of industrial accidents, but which came to be used more for polio victims. They came up with the idea of bellows driven by an electric motor within an airtight tank. The motion of the bellows pushes air in and out of the tank, creating negative and positive pressure. The negative pressure contracts the patient's diaphragm, causing inhalation while the positive pressure expands the diaphragm, causing exhalation. Their US 1906453, filed in 1931, was nicknamed by Drinker the "Iron lung".

Polio outbreaks rose to a peak in 1952, but thankfully Jonas Salk invented the polio vaccine which was used from 1956. He was asked about the patent for it and said "There is no patent. Could you patent the Sun?" No new cases need the iron

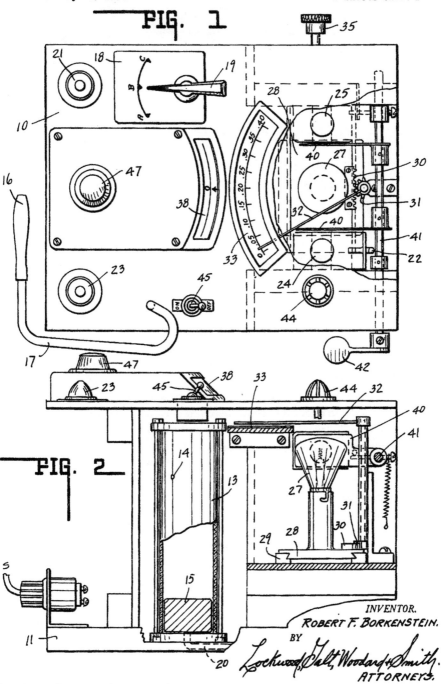

FIG. 1

FIG. 2

INVENTOR.
ROBERT F. BORKENSTEIN.

BY

Lockwood, Galt, Woodard & Smith.
ATTORNEYS.

Breathalyzer® machine (US 2824789)

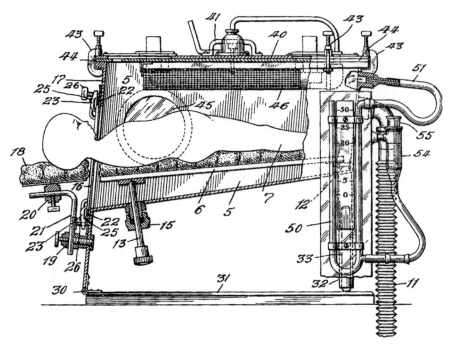

Iron lung (US 1906453)

lung, which anyway was largely replaced with portable respirators in the 1960s, but it has been estimated that a hundred Americans still depend on (probably improved versions) of it. Typically it is used to sleep in, and at intervals during the day. John Goodman of San Diego has used one since 1950. He calls it "an embracing and tireless old friend" and says "being in the iron lung is like being in heaven. It's the only time I can take a really deep breath." Nevertheless many working in the field saw it as a negative idea, as it did not cure, but rather alleviated suffering.

Some little things seem so handy that you wonder how you ever coped before. For example there are Q-Tips®, which were invented in the 1920s by Polish-born Leo Gerstenzang of Long Beach, NY. He had founded a company making baby accessories and one night watched his wife bathing their baby daughter. She put a piece of cotton on a toothpick to clean her baby's ears. He thought how dangerous this was, and wondered about mechanizing the process of making a swab. After years of work he filed in 1927 for US 1721815 for the method of making the product. The huge machine made secure, medically treated double-ended swabs. The process was completely automatic except for feeding the sticks in and adding the cotton to the ends as they were twirled around. At the end, it dropped the swabs into trays and sealed them with cellophane. Originally he called them Baby Gays, but soon changed the name to Q-Tips®.

Another familiar product came from Herbert Lapidus of Ridgefield, CT, who invented the Odor-eater® insole, applying for US 3842519 in 1973. They are simply foam latex inner soles impregnated with activated charcoal, which absorbs the odors. The patent was assigned to Combe Inc. of White Plains, NY, who continue to sell the product. Renfro has made an agreement with Combe to make socks along the same lines, states a January 2002 press release which said that Combe had been a "leader in odor control for over 25 years".

Plastic surgery is now widely accepted for cosmetic reasons as well as to treat disfigurement. Besides "nips and tucks" there are breast implants. General Electric first developed practical silicones, the material that was initially used. Corning Glass, who had cooperated with them, decided to set up on their own in the area. As a glass company they lacked the chemical expertise to develop silicone polymers, or to manufacture them, so they formed a partnership with Dow Chemicals in 1943, and formed Dow Corning. The company offered silicone materials to two plastic surgeons, Thomas Cronin and Frank Gerow of the University of Texas at Houston who developed the first silicone breast implant, patented in 1963 as US 3293663, a "Surgically implantable human breast prothesis". Previously, sponges had been used but these hardened and looked and felt less natural after a period of time. Body tissue also tended to invade the sponge, creating uncomfortable scars. The silicone rubber sac was filled with silicone gel. Where it met the skin a corrugated Dacron® polyester mesh (7 and 9) and "daubs" of cement (10) kept the prothesis in place.

During the 1960s the company improved the tactile properties of their breast implant device. They eliminated a seam and made the outer sac thinner. Saline-filled breast implants were the next improvement. A problem was that the Food and Drugs Administration (FDA) was only responsible for regulating drugs. In 1976 Congress expanded the FDA's responsibility to include regulation of medical devices such as breast implants which were already on the market. As no new money was provided to monitor these new responsibilities most devices (including breast implants) were put in Class II, requiring general controls and performance standards, but no detailed studies. In 1981 a San Francisco attorney successfully sued Dow Corning on behalf of a patient with rheumatoid arthritis. The attorney had read the part of the Medical Device Amendments which states that devices must be proved safe and effective before being released for use. This wording places the burden of proof entirely on the manufacturer if a patient develops a problem and alleges that the device created that problem. Since no one knows what causes rheumatoid arthritis, the manufacturer could not prove that the breast implant did not cause it. The attorney won $3 million for his client.

The FDA responded by proposing to place breast implants in class III, requiring stringent testing. In 1988, after Dow Corning had lost several more lawsuits, the FDA finally reclassified breast implants into class III. In 1992, the FDA placed silicone gel-filled implants under restriction, for research use only, and left

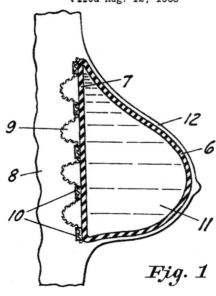

Dec. 27, 1966 T. D. CRONIN 3,293,663

SURGICALLY IMPLANTABLE HUMAN BREAST PROSTHESIS

Filed Aug. 12, 1963

Fig. 1

Fig. 3

Fig. 2

Fig. 4

INVENTOR.
THOMAS D. CRONIN
BY Robert F. Fleming Jr
ATTORNEY

Silicone breast implant (US 3293663)

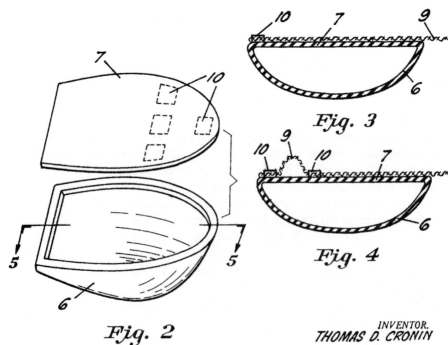

saline-filled implants on the market in the same "not approved but also not restricted" status that had been the case since 1976. In 2000 after examining data the FDA approved saline-filled breast implants for general use. In 2001 there were over 219,000 breast augmentations. There were also 275,000 liposuction operations in what is clearly a booming market. There are many patents for surgical instruments applicable to this field of surgery, with "Power-assisted liposuction instrument with cauterizing cannula assembly" being a typical title.

For many years Kotex® was the leading sanitary towel brand. Its origins are in cellucotton, a woodpulp product marketed from 1914 by Kimberley-Clark which was five times as absorbent as cotton. It was better at controlling infection and only cost half the price. It was supplied as bandages during World War I when some nurses in France realized that they could use them as improvised sanitary towels. Some of them asked the company to sell them for that purpose. At the end of the war the surplus stock was bought back from Europe and the company was faced with the problem of disposing of the stockpile. To avoid a possible backlash, Kimberley-Clark created a second company to market the new product from 1921. Their advertising company suggested it be called Kotex®, for cotton texture (although it was not), but it did not suggest what the product was for. The early versions were huge, 9 by 3 inches in size, and were secured by safety pins to undergarments.

Sales were initially poor as the advertising was vague and many stores refused to stock it. From 1923 a new effort featured the slogan, "The safe solution to women's greatest hygiene problem" and was signed by a real nurse, who was actively involved. Women began to ask nurses for the product. Sales were still handicapped by shyness in an era when goods were behind counters and customers had to ask for what they wanted. Retailers were asked to place the large, unmarked packages in prominent places so that a specific request would not have to be made. Self-service displays enabled women to drop 50 cents into a slot and to leave without asking sales staff. By the late 1920s the idea was so accepted that there was a price war with rival brands, and a struggle for market share has continued ever since.

Tampons as a manufactured product owe their origin to Earle Haas. He was a country general practitioner who moved to Denver in 1928. He invented a flexible ring for a contraceptive diaphragm, from which he made $50,000 by selling the patent, sold real estate and was president of a company that manufactured antiseptics. Haas got the idea for his tampon from a friend in California who used a sponge internally to absorb menstrual flow. He developed a plug of cotton which was inserted by a cardboard tube applicator. A string enabled the cotton to be removed for disposal. It was the applicator that was new about the product, ensuring a monopoly for many years. The patent says that it was "designed for simplicity, economy and efficiency". In 1931 he applied for what became US 1926900, a "Catamenial device" (from the Greek word for monthly). The trademark Tampax® was adopted, from the word tampon, which means plug, and (vaginal)

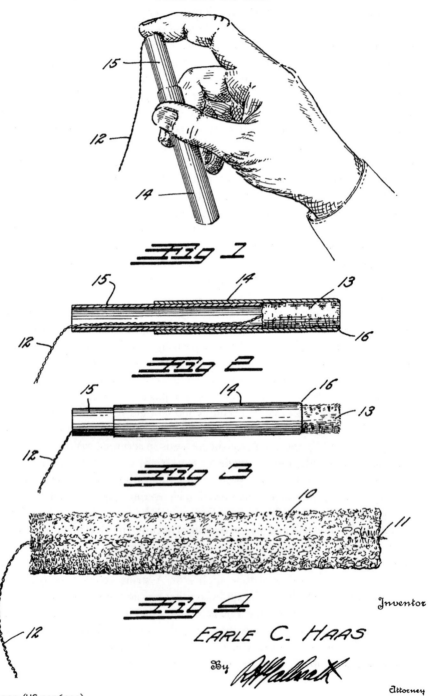

Fig 1

Fig 2

Fig 3

Fig 4

Inventor

EARLE C. HAAS

By

Attorney

Tampon (US 1926900)

pack. Haas' wife, a nurse, tested the product as did other nurses. The Depression was a difficult time to start a new business, and after failing to get companies such as Johnson & Johnson interested, Haas sold his patent and trademark rights in 1933 for $32,000 to a Denver businesswoman, Gertrude Tenderich, a German immigrant. At first she made the product at her home using a sewing machine and a hand-operated compressor. This compressor was then improved by Joseph Voss, her brother, which speeded up production to over 1,000 an hour.

Meanwhile, as with Kotex®, the product was having sales problems, and was solved in much the same way. Stores were shy about displaying the product in its blue and white box, and one trick when visiting drugstores was to ask for a glass of water and to then drop the product in to show how absorbent it was. From 1936 sales rose following a new campaign using full page color advertisements in magazines which included mail-in coupons. The wording of the slogans was crucial, including "In a man's job there's no time for 'not so good days'" and "A man's job doesn't allow for feminine disabilities". With World War II it was claimed that "War work won't wait. . . ." and that the product gave "assured effectiveness in action", and women working in war industries began to identify with it. By the 1950s they had 90% of the market, and the company still has half of the world market. There have of course been improvements, such as moving from cardboard to plastic for the applicator in the early 1980s.

Many assume that exercise equipment is fairly new. In fact from the 1870s half a dozen inventions annually were coming out in the field, many of which look very peculiar. The first adjustable barbell dates from 1911 with Alan Calvert of Philadelphia and his US 1044018. Weights could be added to the spheres. He did, however, also retain the practice dating from the 1890s of allowing extra weights (he suggested shot) to be put into the barbell itself. Weights were pressed while lying on the floor as it was not until 1935 that someone thought of using benches. Inclined benches date from 1953, when one was jammed into a door jamb at an Oakland, CA, gym.

The biggest revolution in exercise equipment, though, is based on the principle of variable resistance. Someone who can lift a certain weight while in one position can lift a much heavier weight while in another position, yet traditional machines only allowed the lifting of the smaller weight to avoid injury. Therefore a lot of training was needed to get results. Arkansan Arthur Jones reasoned that it would be better to make the weight itself vary so that a suitable weight was available at each position. In the late 1930s he tried attaching chains to barbells so that a higher lift would raise the chain off the floor (and hence add weight). He took up flying, and became interested in wildlife, but continued to think about the problem. One night when living in what is now Zimbabwe he woke up at 2 a.m. with the idea that he was convinced would change exercise equipment. He phoned up one of his employees and told him to get a piece of paper and pencil, and for the next 45 minutes dictated to him the dimensions of a component. The employee was expected to make it immediately and to attach it to the existing machine. The next morning Jones realized that it did not work—but looking at it he saw why. He was going to make the motion follow a curve rather than the barbell's up and down direction.

He developed a cam gear shaped like a spiral-shelled nautilus mollusk or a stretched out wheel. Better results are possible from 90 minutes of well-targeted exercise per week than from 20 hours of conventional weightlifting. Ariel Dynamics of Trabuco Canyon, CA, however, claims on its website that among its founder Gideon Ariel's patents is "one on variable resistance which is the basis of most modern exercise equipment", and this seems to be US 4256302 from 1976 which illustrates the variable resistance concept, as well as being a nice drawing of a leg exerciser.

Jones was very insistent that the weights should be increased gradually so that every time you finish a set of exercises you are exhausted. The first machines were developed in a Florida garage on a shoestring budget and were handmade. They became available from 1970. Sales slowly increased by word of mouth, helped by Jones' articles in *Iron man* magazine about his ideas in exchange for which he received free advertisements. A network of thousands of Nautilus gyms grew up. Nautilus also designed the first neck machines from 1975 which are needed to strengthen the muscles and hence prevent injuries. The company did not introduce computerized machines until 1985, 8 years late, due to extensive designing by Jones himself. In 1986 Jones sold Nautilus and launched MedX, introducing the first machines capable of accurately measuring changes in strength, muscular endurance and range of motion. This includes the MedX Lumbar Extension machine, for rehabilitating those suffering from low back pain. It has been said that this is a medical complaint second only to the common cold or headache.

A well known type of machine in the exercise industry is the Stairmaster®. In 1986 Lanny Potts of Tulsa, OK, filed for what was later reissued with corrections as US Re 34959. It was the first workable, compact stair climber. An important feature was that it was not symmetrical. If you pushed down with one foot there was not an equal response with the other pedal coming back, as this could be dangerous.

A major contribution to health care comes of course from medicines developed and sold by pharmaceutical companies. In 2000 the US pharmaceutical market was $152 billion, which is about $548 per person. This represents nearly half of all sales in the world. The cynic could say that pharmaceutical companies make their money by selling medicines and not cures.

Originally the pharmaceutical market consisted of remedies of doubtful efficacy made up in the back room. As placebos they may well have had some value to the patient. Again and again advertising was vital in promoting a new product, or retaining customer loyalty, and trademarks were a vital element in this. Many early trademarks depicted in the Patent Office's official gazette consisted of a depiction

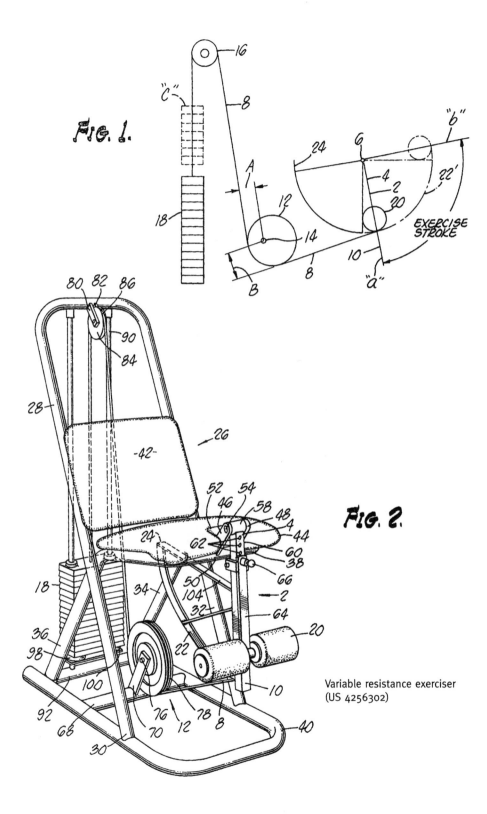

Fig. 1.

Fig. 2.

Variable resistance exerciser
(US 4256302)

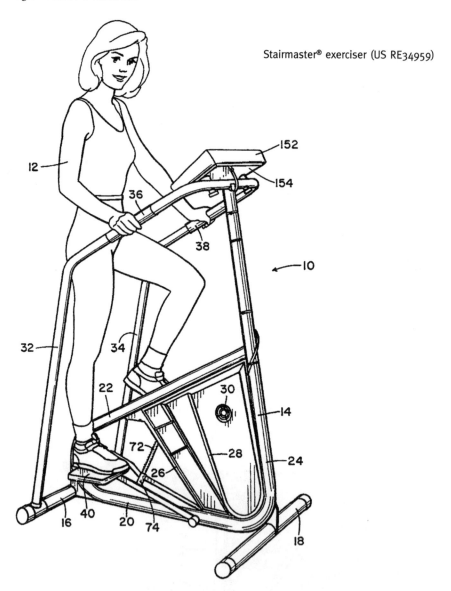

Stairmaster® exerciser (US RE34959)

of the retailer or originator of the potion, often with a signature below to add extra authenticity, while others were odd for other reasons, as shown opposite.

The most successful of these early preparations was probably the famous Dr Williams' Pink Pills for Pale People, which was a Canadian remedy registered as a trademark by a Dr Jackson in 1887. With the exception of Listerine®, marketed from 1879 but not sold as a mouthwash until the 1920s, the first *useful* remedy to sell in large quantities which continues to sell well today is probably aspirin, which

26,406. COUGH-DROPS. CATHERINE DYBALL, Omaha, Nebr. Filed Feb. 18, 1895.

Essential feature.—The representation of two brownies upon either side of a black panel bearing the words "LA GRIPPE." Used since September, 1894.

26,407. CATARRH CURE. GEORGE W. FISHER, Baltimore, Md. Filed Feb. 26, 1895.

Essential feature.—A picture of a man's head and shoulders with handkerchief held before the nose from which discharge is flying, the head thrown backward, mouth open, and the face distorted with pain. Used since December 8, 1893.

26,408. REMEDY FOR EXCESSIVE PERSPIRATION. BENJAMIN F. FRITTS, Chattanooga, Tenn. Filed Mar. 23, 1895.

ANTI-FUT-SWET

Essential feature.—The compound word "ANTI-FUT-SWET." Used since December 1, 1894.

26,409. SPECIFIC FOR SEASICKNESS. YANATAS, LIMITED, London, England. Filed Mar. 8. 1895.

YANATAS

Essential feature.—The word "YANATAS." Used since August 23, 1894.

Early medical trademarks

was marketed by Bayer from 1899. Aspirin can be bought in many guises, including Alka-Seltzer® with its reassuring fizz. It has a "buffer" in it which absorbs excess stomach acid and helps the bloodstream to absorb the medicine quickly (Champagne also quickly affects the consumer, again by using bubbles). It came about when there was an epidemic of influenza in December 1928 affecting half the American population. A newspaper editor in Elkhart, IN, tried using a combination of aspirin and bicarbonate of soda which did the same job as the present remedy. It took 2 years of research for the Dr Miles Laboratories in the same town to make a suitable tablet that could be used in water. Launched in 1930, its continued popularity shows the power of heavy advertising, with early radio advertisements saying "Listen to it fizz !" as the product did just that. It was also boosted by being used to cure hangovers with the repeal of Prohibition in 1934.

Aspirin itself, plain and simple, can be bought as a non-branded product for a lower price than a trademarked product. That is because it is no longer protected by a patent. After expiry a popular trademark can be retained by the original company to keep the product popular but most consumers prefer the rivals' lower price. In the case of aspirin it lost its trademark in a 1921 court case on the grounds that it was "generic" (like a noun) although it survives in other jurisdictions. Interestingly enough, in 1996 Bayer tried to register Aspirin as a trademark for cut flowers, presumably thinking of the use of the product to keep them alive for longer. The difference in price with other companies' "generic" versions of the drug once a drug is off-patent is often huge, down to say a fifth of the original price. The companies argue that they need the prices to pay for the research into numerous possible drugs, few of which ever make it to the marketplace, and testing. There are no price controls, and some go to Canada or Mexico to buy supplies of what they need.

American patents being filed now have 20 years protection from the date of filing (formerly 17 years from the grant (publication) of the patent). Either way there is a fixed term and it was argued that this was unfair to manufacturers if they had to wait for permission by the FDA (Food and Drugs Administration) to market the drug while the patent was already in force. The 1984 Waxman-Hatch Act allowed for up to 5 years' extension to the term to allow for this. It also provided the present laws for "generic" drugs, which have to be bio-equivalent to the original drug. In 2000 about 42% of all prescriptions in the United States were filled by generic drugs, according to the FDA, but they make up only 18% of the market by value, which shows their cheapness.

Consumers would like cheap yet reliable generics. The way they actually get to the market is complex, and may seem strange. To get a drug approved as safe and ready to launch at the time the patent expires, a generic drug-maker must file for an Abbreviated New Drug Application (ANDA). The company affected must be notified and has 45 days to initiate litigation for patent infringement. If the patent holder brings a suit within the 45 day period, the FDA delays the approval for a period of 30 months or until the resolution of the litigation, whichever comes first. Of course, since there is every incentive for the branded drug company to litigate, that tends to be what happens. The need for speed is driven in part by the 180 day exclusivity granted by the FDA to the first company to file an ANDA for a given drug. Since the 180 days represents an opportunity to win market share and keep prices (relatively) high, this period often means the bulk of the generic manufacturer's profits, as later entries by other generics drive prices down.

Overall, drugs with annual sales of about $45 billion are set to go off patent between 2001 and 2005. Even a few days' delay in releasing a generic drug is useful for the original company if a million dollars' worth is sold daily, and not surprisingly the share prices of pharmaceuticals can go up or down according to a company's current litigation. The situation is even more complex than it might seem as there is not just the patent for its preparation but also separate patents for making it up into a tablet, syrup and so on. Because of the potential value of the patents, monitoring and analysis as well as litigation is very intense in the area.

Some well-known drugs include Paracetamol® or acetaminophen by Warner-Lambert in 1958 with US2998450; Valium® or diazepam, "Mother's little helper", filed in 1960 as US 3371085 by Hoffman La Roche; and Prozac® or fluoxetine which was filed by Eli Lilly in 1974 with US 4314081. The illustration on page 136 is from the Prozac® patent, which makes dull reading for those who are not chemists.

In the future there may be personalized medicines, a science known as pharmacogenetics, which will be tailor-made by checking your personal DNA sequence. While many will benefit from treatments resulting from research into DNA and related fields, the question of patenting DNA, and property, can lead to interesting (and to many, bizarre) arguments. In 1976 John Moore visited the University of California, Los Angeles to get medical treatment for his hairy cell leukaemia. He was told that he was cured, but they kept asking him to come back and give more tissue samples. Moore finally became suspicious and discovered that the university had filed in 1981 for a patent which was published as US 4438032, with the title "Unique T-lymphocyte line and products derived therefrom". The cells had unusual (and financially lucrative) properties. Angry, Moore sued the five bodies concerned for breach of trust, lack of informed consent, and "conversion", defined by the court as "the wrongful exercise of ownership over personal property belonging to another". The defence was that Moore's cells were indeed their property. In 1990 the California Supreme Court held that, since Moore had no property rights in his body, he had no rights to the profits. The question of property rights in your own body is a state matter, and some states have passed laws to clarify individuals' rights over their own bodies.

Startling examples of engineering technology can also be used in attempts to improve our health. Something out of a B-movie has been patented by Interval Research of Palo Alto, CA, as US 5638832, which dates from 1995. The

4,314,081

1

ARYLOXYPHENYLPROPYLAMINES

BACKGROUND OF THE INVENTION

Tertiary 2-phenoxy-2-phenylethylamines constitute the subject matter of U.S. Pat. No. 3,106,564. The compounds are said to be useful pharmacological agents exhibiting activity on the central nervous system including useful application as analeptic agents without significant effect on respiration. The compounds are also said to have a high order of activity as antihistaminic and anticholinergic agents. Several tertiary 3-phenoxy-3-phenylpropylamines and quaternary ammonium compounds are disclosed in *J. Pharmaceutical Society*, Japan, 93, 508–519, 1144–53, 1154–61 (1973). The compounds are said to be mydriatic agents.

Secondary and primary 3-aryloxy-3-phenylpropylamines have not hitherto been known.

SUMMARY OF THE INVENTION

This invention provides 3-aryloxy-3-phenylpropylamines of the formula:

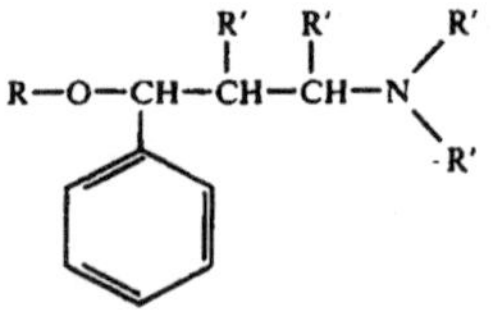

N,N-dimethyl 3-(α bromide
N,N-dimethyl 3 propylamine iodide
3-(2'-methyl-4',5'-c propylamine nitrate
3-(p-t-butylpheno N-methyl methylpropylamine
3-(2',4'-dichlroroph mine citrate
N,N-dimethyl propylamine maleat
N-methyl 3-(p-tr
N,N-dimethy propylamine 2,4
3-(o-ethylphenc gen phosphate
N-methyl 3-(2'-cı, nyl-2-methylpropylan
N,N-dimethyl 3· nyl-propylamine su
N,N-dimethyl propylamine pheny
N,N-dimethyl propylamine β-phe
N-methyl 3-(p- propiolate
N-methyl 3-(3-n· mine decanoate
Also included w
the pharmaceutical

From the Prozac® preparation patent (US 4314081)

"Programmable subcutaneous visible implant" is a tattoo implant in the human arm. Previously there had been simple devices for livestock, but this was a more sophisticated device. A small LCD (liquid crystal display) is inserted just beneath the skin on the wrist, where most people wear a watch. Because human skin is partially transparent, the display is clearly visible. The display is connected to a control chip and a small battery provides power, both being implanted beneath the skin. Implanting is an outpatient operation and the battery can be recharged by holding the wrist near a charger. The tattoo is meant to contain biosensors to monitor temperature and blood pressure, and to display these readings on the watch.

The company was started to think creatively, with ideas that might not be reality until 5 to 20 years down the road. It started up in 1992, one of its founders, Paul Allen (who had earlier been a cofounder of Microsoft) committing $100 million. The company folded in 2000. It had been hoped that it would emulate the

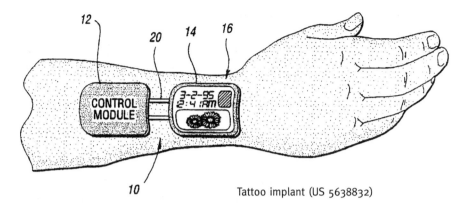

Tattoo implant (US 5638832)

creativeness that made the PC. The company had a reputation for secrecy, but over 50 patents explained many of its ideas.

Not everybody wants a tattoo on their forearm. A more attractive idea may be wearing clothes which monitor your health and transmit data, perhaps down a phone line to medical staff far away. One example of attempts along these lines is the application US 2002/0169387, a "Wearable body fat sensor" by U.S. Phillips. Special shoes are worn which weigh the patient, as well as a tight-fitting costume. A step further was taken by US 6248064 from 1998 by ineedmd.com of Great Neck, NY. The "Tele-diagnostic device" consists of gloves which provide medical data to a remote location. They suggest that the data would include readouts from an electrocardiograph, blood pressure and pulse, oxygen levels in the blood, temperature and a stethoscope. Some help could be given as well, as the glove would include a defibrillator (in case of a cardiac arrest). Such ideas are likely to be reality in the not too distant future.

Among other kinds of health products, Vaseline® was invented by Robert Chesebrough, who was born in London, England but to American parents. He worked as a chemist and by selling kerosene and his business was affected by the first oil strike in 1859 at Titusville, PA. He went to have a look and workers told him that the "rod wax"—oil which clung to the rods of the pumps—cured cuts and burns. It was a nuisance as it also gummed up the machinery. He took some of it back to his Brooklyn lab and worked on the substance. He knew that the Egyptians used oil as a healing balm. He cut or burnt himself to try out each formulation. He came up with a clear jelly (which it is if you heat it), filtered through "bone-black". It was not until 1872 that he got his US 127568, which says "Vaseline is a thick, oily, pasty substance" and which recommended it from experience as a hair lotion. His name Vaseline® came from the German *wasser* for water and *elaion*, the Greek for olive oil. He went around New York City in a horse and buggy handing out free samples, which is said to have been the first "give-away" campaign. In 1881 he sold out to his supplier, Standard Oil. Chesebrough himself swallowed a spoonful a day, and had himself rubbed in it to cure his pleurisy. He lived to be 96.

The original formulation for Mum® deodorant was invented in 1888 by a mysterious man from Philadelphia who may have been Seraph Deal (at least, it was he who registered the trademark in 1908, claiming its use from 1888). This is thought to have been the world's first anti-perspirant. It was a cream that could be applied from the jar with your fingertips. In 1931 the pharmaceutical company Bristol Myers acquired ownership. It was their only such product, and after a couple of decades its sales began to fall behind rival Arrid®. The alternatives to cream at the time were lotions or sprays. An outside inventor developed for them the roll-on concept, inspired by the rotational action of a ballpoint pen. The product was tried out in six test markets in 1951 as Mum Rollette, but was unsuccessful, as the ingredients damaged the plastic used in the rolling ball. Bristol Myers still liked the principle and decided to persevere. They tried another 480 combinations of plastics before another company managed to solve the problem for them.

In 1955 the product was relaunched as Ban® roll-on and was an immediate success. Within a couple of years it was the leading anti-perspirant, helped by a major advertising campaign on television. This featured a woman demonstrating how the product was used while a male voice delivered the sales pitch. A fresh challenge appeared in the mid 1960s, aerosol sprays, before roll-ons surged back in the late 1970s as aerosols began to lose their appeal on environmental grounds. The aerosols had in fact had a problem when first used: they could stain clothes. The solution was found to be the incorporation of "volatile silicones" in the formulation. Aerosols themselves date back to research in 1941 by the Department of Agriculture to control malaria, although the first "clog-free" valve dates back to 1953 and Robert Alplanal. Unfortunately the effects of chlorofluorocarbon or CFC was not appreciated at the time.

Benjamin Green, a Miami Beach pharmacist, had noticed that sunbathers were using improvised substances on their skin. He experimented by cooking cocoa butter in his wife's coffee pot and testing it on his (bald) head. In 1944 he launched his new product, Coppertone® Suntan Cream. The bottles showed an Indian Chief and had the slogan "Don't be a paleface". This, of course, was when people liked a tan rather than being afraid of what it would do to their skin. The famous "Little Miss Coppertone" with her bathing costume being tugged down by a dog, showing the contrast in skin color, dates from a 1953 campaign. It was on a giant billboard for many years and was nicknamed "the moon over Miami". When the company tried to remove it in 1991 there was a protest about the loss of this piece of Americana, and so the billboard remained.

Since about 1998, the phrase super soap has sometimes been used to refer to the recent crop of anti-bacterial soaps that are crowding regular soaps off the shelves. Anti-bacterials are designed to kill germs, but supersoap is different. At the general meeting of the American Society for Microbiology, held in May 2002, scientists from Colgate-Palmolive "unveiled" in a paper something called Microbial Anti-attachment Technology (MAT). It was patented as US 6165443 which had been published 18 months before, so it was hardly new. It is based on three common cosmetic ingredients: petrolatum (otherwise known as Vaseline®), dimethicone, and polyquaternium. MAT coats the skin with a super-thin film that repels rather than kills bacteria. Their press release said that MAT gives the skin a "Teflon-like" property. The scientists claimed that "bacteria attachment" was cut in half using a MAT-based soap. Simply frequently washing your hands in water has been shown to very be effective as well.

All good things must eventually come to an end. One way of predicting this was invented by David Kendrick of Berkshire, NY, who in 1991 applied for his "Life expectancy timepiece" with US 5031161. You program in about 10 facts and it counts down to the expected date of your demise, displayed as "time remaining". He felt that it would be "advantageous to monitor the probable remaining time left in one's life on a yearly, hourly, and even seconds basis". In the drawing the wearer has on average 37 years, 100 days, 21 hours, 4 minutes and 42 seconds remaining. To assist the owner of the watch, the patent contains standard actuarial tables of life expectancy together with suggested pluses and minuses according to where you live, weight, smoking habits and so on. A placid or good-natured disposition gives you between 1 and 5 years, according to your choice, while being married adds 5 years.

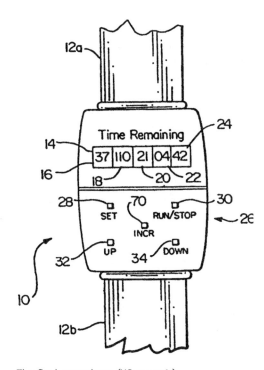

The final countdown (US 5031161)

The open road

H AVING your own car with the freedom of the open road allows you to go wher-
ever you want. In such a big country an easy way of getting round is vital. The
suburbs originally grew with the provision of mass transport, especially street cars
and other light railways, but cars have clearly dominated for a long time. The car
is the owner's own environment, and can be like a home in many ways, with personal touches. This has been helped along by such innovations as enclosing the car
(in 1916 only 2% of manufactured America cars were enclosed, but by 1926 67%
were) and providing air-conditioning from the 1950s. Public transport is imper-
sonal, inconvenient, and often dirty, and availability is often poor with the growing number of cars making it uneconomical to run. Sometimes, it has been said,
public transport was deliberately closed down by interests which intended to make
money out of highways. As for walking to your destination . . . well, enough said.
Eighty-seven per cent of Americans get to work in a car, while only 3% walk.
Whether all this is sustainable is another matter.

The early innovations in making what was, significantly enough, initially the
"horseless carriage" came from Europe, mainly Germany and France. It was truly
the beginning of a transport revolution. Cars meant lost jobs, such as for saddle
maker William Hoover, who had to turn to manufacturing his brother in law's new
vacuum cleaner. But many more jobs were created in such places as filling stations,
car washes, drive-in movies and drive-in restaurants. This was apart from actually
manufacturing the cars and spare parts, and in the repair shops.

By 1909 over half of all American car production was in Michigan. The reason
for this was that the state had a tradition of machine shops and a knowledge of gasoline engines through making marine engines. Iron and coal could easily be shipped
by water. Also, a number of the pioneer engineers were born or worked in Michigan.
By 1914 the headquarters of the "Big Three", Ford, Chrysler and General Motors
were all in Detroit, although it was not until 1932 that their combined production
was greater than all the dwindling band of independents put together. In 1900
American manufacturers built just over 4,000 cars. By 1920 it had risen to close to 2
million and then to a peak of 4.4 million by 1929, when it fell to a low of 1.1 million
in 1932, not to rise to a new peak until 1949. More revealing than the actual numbers being built (or imported) is the ratio of population to the number of registrations, which was 13 to 1 in 1920, 2.9 to 1 in 1960 and 1.9 to 1 in 2000. Already by the
1920s, following the devastation caused by World War I in Europe, America was
overwhelmingly ahead of all other countries in car ownership.

Few realize that there was a struggle between electric- and gasoline-powered
cars in the early years, and it was not obvious at first which side would triumph.

The Duryea brothers were firmly on the side of gasoline, and are widely credited with the first American gasoline-powered car that actually moved, if only for 200 feet. This was in 1893. They formed the Duryea Motor Wagon Company in 1896, which was the first American company to make gasoline-powered cars, with no fewer than 13 in their first production run. The illustrated US 572051 is by James Duryea of Springfield, MA. The invention was for improving the power transmission from the crankshaft to the driving-axle and also for improving the means of varying the speed and for reversing.

Besides the actual construction of the car, many accessories were gradually invented or improved. One winter Mary Anderson of Birmingham, AL, visited New York City where she rode on a streetcar. She could not help noticing that the motorman could hardly see through the ice and snow accumulating on the windshield—for wipers did not exist. He would occasionally stop and go out to remove the ice by hand, or would peer out of the side window. She began to make sketches, and in 1903 applied for what was to become US 743801. It was non-automatic, as you had to use a handle inside the vehicle. The spring-loaded arm returned after making a sweep. By 1916 nearly all new cars had some sort of windshield wiper.

Early motor horns sounded too soft and another important, if soon to be superseded, advance was invented by Alabaman Miller Reese Hutchison in about 1907. He was an assistant to Edison when "I was driving in Newark one evening. It was raining, foggy, and the streets were wet and slippery. A man darted out immediately in front of my car. I applied the brakes and sounded my horn, which gave out a little musical note somewhat resembling an angel's harp. The pedestrian almost joined an angelic choir. So then and there I decided that cars should have a horn that startled, shocked and repelled instead of one with a pleasing sound." The result was the famous "Ah-OOGA" klaxon operated by squeezing a rubber bulb.

Another important invention was when Charles Kettering of Dayton, OH, devised an integrated system (based on his experience making cash registers) incorporating an electrical ignition system, a self-starter for the engine, and the first practical engine-driven generator. His US 1150523, filed in 1911, spoke of previously using a foot pedal and a crank. There was no need now to swing a rotor manually at the front, and his invention was a great success. Electric self starters had already started to appear in manufactured cars, and this gave independence to those who could not physically have cranked the cars by themselves. Women felt more liberated as a result.

Enclosed cars were available from about 1900 on but as they cost about 20% more than open vehicles few were bought. To protect passengers in open vehicles, car accessory companies sold folding cape and canopy tops. The closed car (or "sedan", after the French horse drawn carriage) became less expensive when Budd cut the manufacturing costs. In 1919, Dodge brought out the first enclosed car with steel frame members and body panels. Nevertheless, the ability to adjust the appearance of your car—to have more choice—by deciding whether you wanted it

J. F. DURYEA.
MOTOR VEHICLE.

No. 572,051.

Patented Nov. 24, 1896.

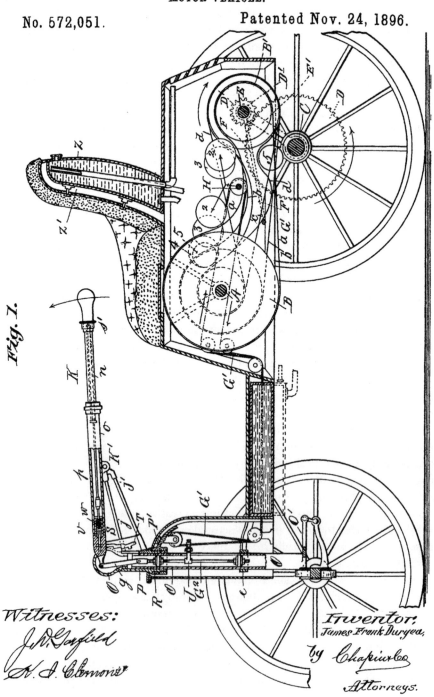

Fig. I.

Witnesses:

J. N. Garfield

N. A. Clemons

Inventor;
James Frank Duryea,

by Chapin & Co
Attorneys.

Duryea car (US 572051)

No. 743,801. PATENTED NOV. 10, 1903.

M. ANDERSON.

WINDOW CLEANING DEVICE.

APPLICATION FILED JUNE 18, 1903.

NO MODEL.

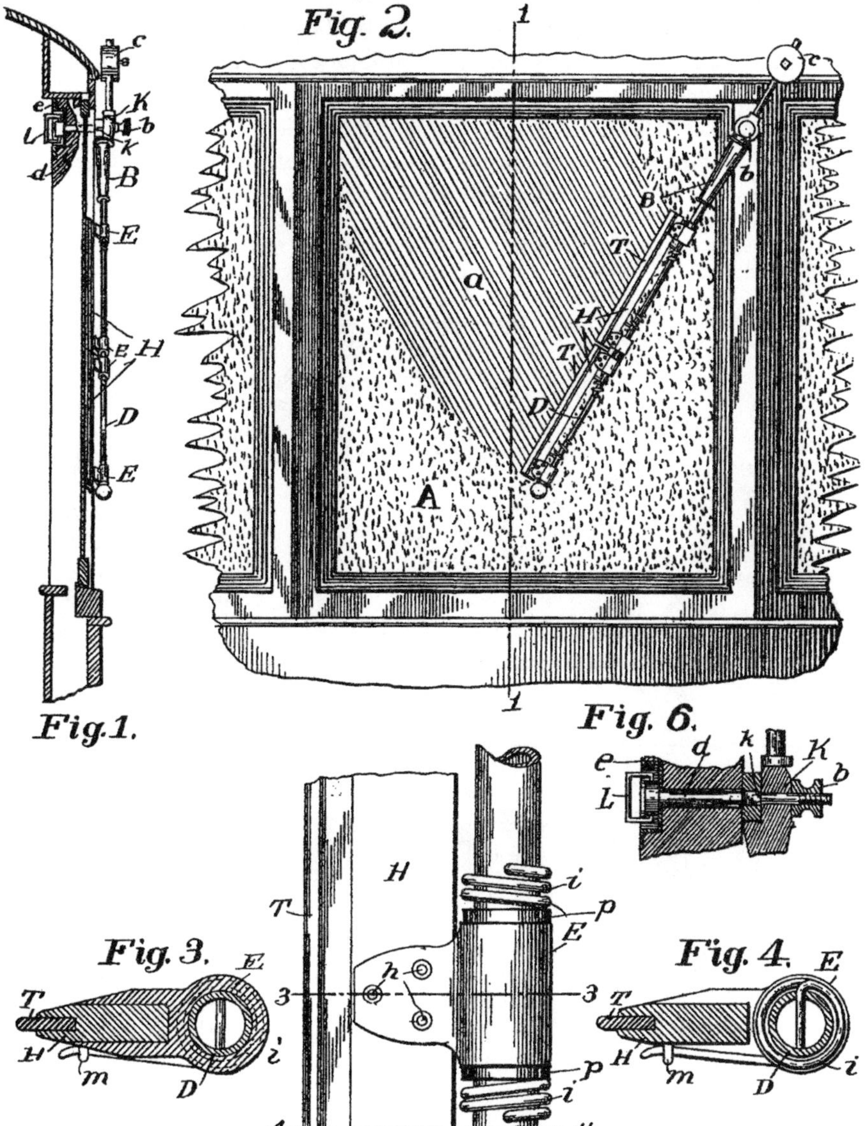
Fig. 1.
Fig. 2.
Fig. 3.
Fig. 4.
Fig. 5.
Fig. 6.

Witnesses
Milton Lenoir
Watts T. Estabrook

Inventor
Mary Anderson
by Shannon C. Hodges
her Attorneys

Windshield wiper (US 743801)

open or closed is surely part of the American Dream. In 1921 John McGuire of Yonkers, NY, applied for US 1433466: a car that could be converted into one of four designs. Within minutes, whatever body was on the chassis could be unbolted, lifted off and replaced with a different body. The number of doors available changed as well. The drawings show his two-door coupé and the four-door sedan. The other models were a two-door (open) roadster and the four-door "touring car with Victoria top", open but with a canopy at the back.

Sadly, McGuire's ingenious idea was not successful, although it might be welcome today. It is Ben Ellerbeck of Salt Lake City who is credited with the hardtop

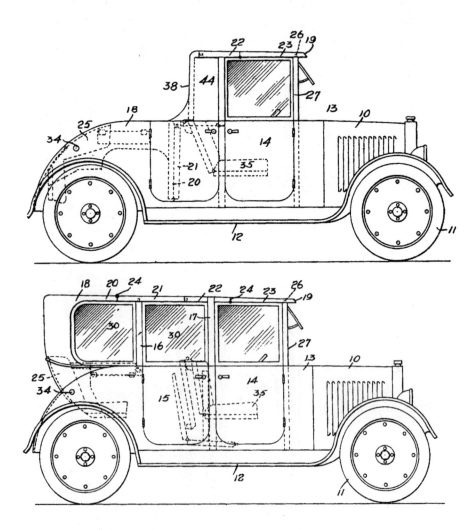

Car converting to four designs (US 1433466)

retractable convertible in 1929, with US 1784279, while convertible models were not offered for sale in any quantity until 1946, by Chrysler.

An essential product that must be incorporated into any car is the car radio. The Galvin Manufacturing Corporation of Chicago is widely credited with the first car radio, with its first patent being US 1959869, filed in 1930 by William Lear. It was for a radio with its controls on a panel attached to the steering column. Cables led to the radio which was fastened to the dashboard. Lear himself was an interesting character, a Missouran who never made it beyond the eighth grade, yet who later went on to achieve his dream of creating a businessman's aircraft with the famous Lear Jet from 1963. There are some earlier claims for car radios. These include a 1924 patent in Australia and a patent for a radio in a Model T Ford in 1914 (which only worked when it was parked), but this was the first to be mass produced. By 1937 the company's car radios had push-button tuning, fine-tuning and tone control, all firsts. The name Motorola® was used for the company's new product to combine the idea of motion and radio. The trademark became so well known that in 1947 the name of the company was changed to Motorola, Inc.

Equally, air conditioning makes driving bearable in hot summers. A rather unwieldy effort was US 2050381, filed by Horace Rogers of Norfolk, VA, in 1936. It was a dual system for heating or cooling where a tank in the roof held water and was pumped back up. The system included a heated foot rail, the supply of water to the radiator if necessary, and spraying water onto the windshield (to cool the interior). Rogers called his mechanism "comparatively simple". The water was likely to be rather warm on a summer's day.

The first production car to be made with air conditioning was the 1940 Packard. The "cooling coil", a large evaporator, was located behind the seat, and the only control was a blower switch. The system was advertised as a "Weather Conditioner" and also filtered pollen and dust from the air. Here too it could be transformed into a heating system by adjusting damper controls located in the trunk. To turn it off you had to open the hood and remove the belt. It was so expensive that between 1940 and 1942 only 1,500 cars were equipped on models such as the Custom Super-Eight. Packard's only patent at the time on the subject is the 1941 US 2344864, a neat system using sulphur dioxide as the cooling agent. In 1957 the Cadillac Eldorado Brougham came with air conditioning as standard, the first time it was offered other than as an option.

The 1940s meant famine followed by feast. No passenger cars were built between January 1942 and July 1945 as the war effort diverted resources, quite apart from the rationing of gasoline (and 35 miles per hour speed limits). With the war over people were yearning to buy new cars. One attempt to satisfy this pent-up demand was made by Preston Tucker. His was the last attempt by an independent car maker to break into mass-produced cars. His car was hailed as "the first completely new car in fifty years" while advertising promised that it was "the car you have been waiting for". Others saw it as a fraud or a pipe dream. His

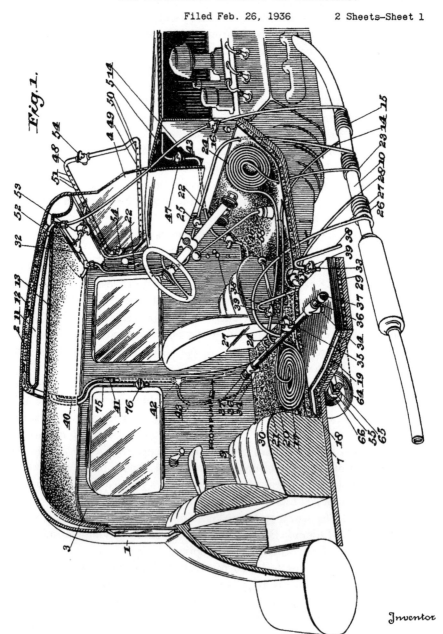

Fig. 1.

Inventor

Horace B. Rogers.

By *Munn, Anderson & Liddy*

Attorney

Car air conditioner (US 2050381)

Tucker Corporation had made money manufacturing his patented gun turrets in the war and was ready for a fresh challenge. US D154192 showed the sleek, elegant lines of his car. US 2465825 was for the instrument panel inside the steering wheel while US 2501796 was for independent wheel suspension. The engine was made of aluminum, there was fuel injection, disc brakes and seat belts. These were innovations only seen on racetracks. There was also a middle headlight that turned with the steering wheel.

Michigan Senator Homer Ferguson and industry lobbyists attacked Tucker, possibly motivated by the thought of the expense of having to supply similar novelties. In 1949 Tucker and his associates were tried on numerous counts of mail fraud, Securities and Exchange Commission regulation violations, and conspiracy to defraud. The prosecution based its entire case on the original "Tin Goose" prototype (so called because it was fashioned from hand-beaten iron), and refused

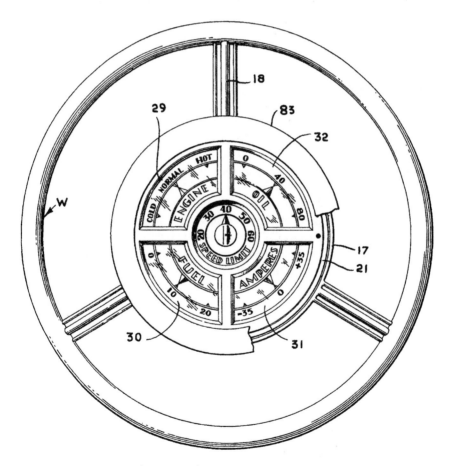

Tucker instrument panel (US 2465825)

to recognize the only 50 production cars to be made. They called witness after witness who, under cross-examination, hurt the prosecution. The jury found the defendants innocent of any attempt to defraud, but the company was so short of money that the remaining assets, including the Tucker cars, were sold for 18 cents in the dollar. Tucker tried again and in 1951 went to Brazil to seek backing for another new car. With the new project almost underway, he was diagnosed with lung cancer and died in 1956.

The war also influenced car styling. Harley Earl was the man who gave us the classic "fins" look. He had joined General Motors in 1927 as a stylist and ran their Styling Division between 1940 and his retirement in 1959. This Californian was a big, quiet man, who dressed in a natty fashion. It was said that "He looked like he could kill you; a near miss would have done". He was always called "Misterearl" by his staff. He designed by making suggestions to subordinates rather than by actually drawing anything himself. Before the war he had changed the look of General Motor cars from a loose assembly of running boards, fenders and mudguards, attached to a big chassis to a lower, wider and (by the 1960s) often much longer car. As far as possible everything was incorporated and virtually sculpted in, with modelling clay being used to tinker with the look. Above all, Earl loved curves: warm curves, sweeping curves, whatever. Even if they did not make the car any faster they looked as if they might. People in the industry began to realize that if you made a car look good it might increase sales.

During the war, Earl's team was invited to a secret showing of the Lockheed P-38 fighter plane (the one with the short fuselage, with the engine pods extending beyond the wings and uniting at the rear, with two oval-shaped tailplanes). They took in what they could from 30 feet. Earl loved the fin-like look of the tailplanes, and it stayed in his mind as a possible new look for cars. He suggested using them over the raised rearlights in the 1948 Cadillac model, but the board rejected it, so he said to his staff "Take that goddam fin off, nobody wants it". The designer refused, so Earl threatened to fire him, and the fin had to be covered up whenever he was in the studio. In the end the board relented and allowed the change.

The "fin" look gradually caught on, and other companies began to follow suit. Ford was the last holdout, but they too went over to the new look in 1957. Gradually the fins became more and more massive, with other features also adopting a new appearance. Earl's own US D180509, from 1956, has a huge fin going down the middle of the trunk as well as bomb-shaped holders for the tail lights, and looked as if it were going to take off. Rockets look more likely as an inspiration than the P-38, but the fancy additions in fact slowed everything up by making the cars heavier and less aerodynamic. Earl was also responsible for the massive grilles and angled fenders with bullet-like braces (US D179125, shown here, typifies this) and the big indentations for the tail "lamp cluster" (as in US D175018). The most extreme form was probably the look of the 1959 Cadillac Fleetwood. Clearly they

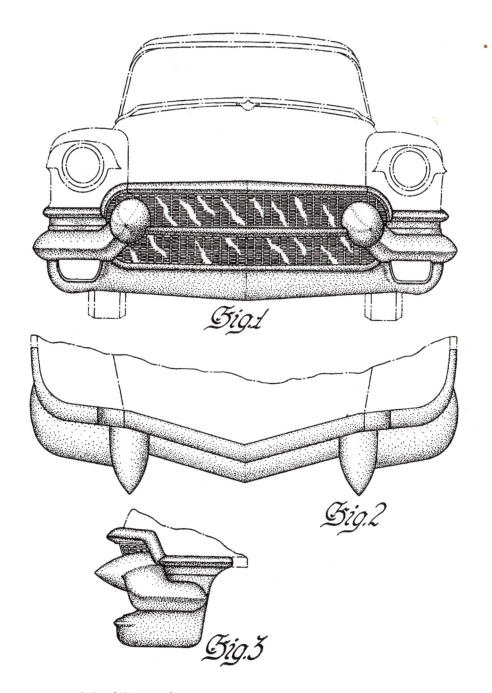

Fig.1

Fig.2

Fig.3

1950s car design (US D179125)

were popular, or people would not have bought the cars. With the 1960s a more "modern" look began to prevail, perhaps encouraged by the failure of the 1957 Ford Edsel, with its heavy lines. Earl's philosophy could be summed up in his "You can design a car so that every time you get in it, it's a relief—you have a vacation for a while."

Of course, even nice looking cars had to be washed. Gradually the idea that car washing meant driving or moving the car through the washing, rather than moving round a stationary car, evolved as adaptations of designs meant for washing railroad carriages. These are easier to construct as carriages are much more likely to have predictable (and long) sides. One of the first patents on that subject was US 1298096 by George Roberts of Chicago, which dates from 1917. The carriages were moved past huge cylinders rotating on a vertical axis which washed the sides. Ingeniously, reversing the carriage through the cylinders allowed drying. The magnificently detailed US 3035293 by Sherman Larson of Detroit, filed in 1956, is of a more recognizable if very compact carwash which also shows typical car styling of the time.

Larson claimed that earlier inventions in the field were bulky and did a poor job, and that soaping and pre-rinsing still had to be done by hand. Rails underneath at (24) guided the car through by hauling on a chain.

Good looking (and clean) cars were all very well, but people were getting killed and injured daily, often because of poor car design. Claire Straith, a Detroit plastic surgeon, was so horrified by what crashes did to faces that he talked to Walter Chrysler. The resulting 1937 Dodge had a smooth dashboard, raised to prevent knee injuries, with recessed knobs. The front seat's back was padded to protect passengers in the rear. There was little interest. In 1949 Nash Motors was the first to offer "lap belt" seat belts in 40,000 of its cars. There was consumer resistance— it suggested driving was unsafe—so they did not do it the following year. From 1955 lap belts did begin to appear, but only as options.

Another way of preventing injury or death to those in cars dates back to the same era. William Haddon was an epidemiologist: someone who used statistics to measure the rate at which people contract illnesses or die. He was a crew-cut, cerebral New Englander who first became known when in 1957 he challenged Daniel Patrick Moynihan (later the Senator, but at the time an aide to Governor Harriman of New York) at a public meeting. He wanted to know what data was being used to assess attempts to reduce traffic injuries. Moynihan had to admit that he was not using any data. Like everyone else he was trying to reduce the number of accidents by controlling speed. They retired to Yezzi's bar across the road, and Moynihan was won over by Haddon's arguments.

Haddon said that crashes were two separate events: a collision between vehicles, followed by a collision between the occupants and the vehicle itself. A metal cage round the car limited problems from the first kind of collision, but something else was needed for the second kind of collision to allow the occupants to decelerate

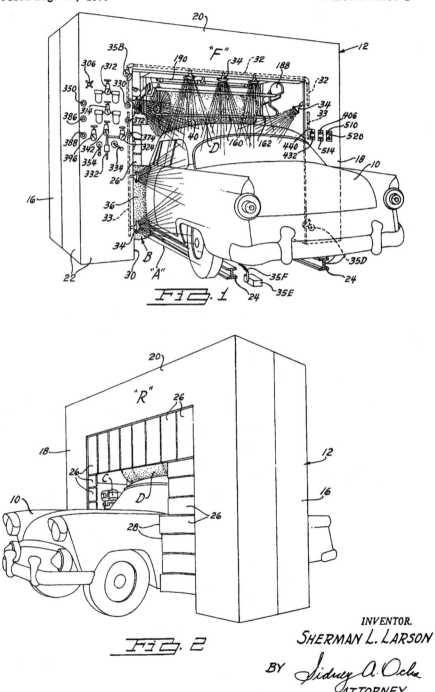

FIG. 1

FIG. 2

INVENTOR.

SHERMAN L. LARSON

BY   Sidney A. Ochs

ATTORNEY

Car wash (US 3035293)

safely. The problem with seat belts was that only 12% of car occupants actually used them. A passive solution that did not require active cooperation was needed. Gradually Haddon and supporters such as Ralph Nader made their influence felt, and the 1966 Motor Vehicle Safety Act required seat belts, energy absorbing steering columns and dashboards, warning flashers and head restraints on newly sold cars. One day in 1968, when Haddon was the first incumbent of the post of Highway Traffic Safety Commissioner, a group of engineers visited him at his office with a suggested passive solution. They called it the People Saver, and we now know it as the air bag.

The concept originated with John Hetrick of Newport, PA, with his US 2649311, filed in 1952. His "Safety cushion assembly for automotive vehicles" was primitive by modern standards. He had once run his car into a ditch, and the impact of abruptly coming to a stop made him wonder if some kind of cushion could protect people in crashes. Hetrick realized that it would have to inflate in a split second—but what could inflate so rapidly? Then he remembered an incident from his time in the Navy in World War II when he was working in a torpedo-maintenance shop. The torpedoes were powered by compressed air, and one day a canvas-covered torpedo accidentally switched itself on. The canvas "shot up into the air, quicker than you could blink an eye", he later recalled. He wrote to car makers and insurance companies when trying to market his new invention. Only one letter was answered: the company replied that they were not interested because people wanted "fancy radios and fancy cars. And I think at the time, they had these fancy fenders. But they were not interested in safety".

Others began to work on the idea. Early models were tested out, with one killing a baboon. There were worries about the impact of setting off what was effectively an explosion inside a car, and air bags now have less power than the first models because of injuries that have occurred to children and smaller adults. The first one that really worked as intended was patented as US 3514124, filed in 1967 by Eaton Yale & Towne, a Cleveland company. As it was not published until 1970 the idea was still confidential when their engineers met Haddon at his office. It was an "energy absorbing panel member which is releasably secured to an interior part of the vehicle" which could be located in the dashboard—or, as shown, in the back of the front seat to protect those in the rear seats. This would be considered a relatively new idea by many. A collision, the patent went on, would act on a sensor (44) and cause a compressed fluid to be ejected from a container. Haddon was ecstatic about what he called a "technological vaccine". The air bag was offered in the 1973 Chevrolet as an option but was only offered as standard in 1988, by Chrysler. The early ones really did have compressed air, but by 1994 gas-powered bags were available.

In the field of cars as in just about any field of American endeavor there is often litigation. What is potentially one of the most lucrative causes of litigation in American history concerns windshield wipers. In 1953 newlywed Bob Kearns

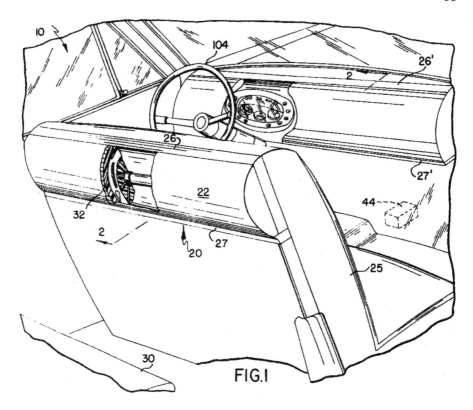

FIG.1

Air bag (US 3514124)

opened a bottle of champagne in his honeymoon suite, only to sustain permanent damage to one of his eyes. That injury made it difficult for him to see through two-speed windshield wipers, and he began to work on an intermittent wiper that would imitate the movement of a blinking eye. He had a number of patents, the key one being US 3351836, which he filed in 1964 from Detroit for the Tann company.

Improved blades interested car makers because wipers help to sell cars. In the late 1950s some models began to have two blades that swept in parallel across the windshield, replacing the single blade that created a huge V in the middle of the windshield. This new two blade system attracted buyers, and by the 1960s every car had them. The next step was intermittent operation, operated by a cheap timer for the wipers. The car companies had developed a mechanical contraption with some 29 moving parts but Kearns used an electric motor with a timer which had four parts, only one of which even moved. It seemed so obvious that the makers thought Kearns' patents would not be valid. He has sued nearly every major car maker for a total of $1.6 billion dollars: $500 million for his lost profits, and more than $1 billion in damages. The first car maker he sued offered him $30 million to settle out of court. Kearns disliked accepting the settlement, as that would suggest that it was acceptable to infringe patents. It took 12 years for the case to go to trial during which his pursuit of justice lost him his wife and broke his health. Eventually Ford and then Chrysler settled for a total of over $30 million. Kearns's offspring have joined Kearns Associates, a company set up specifically to litigate patent-infringement claims. Its corporate offices are conveniently located across the street from the federal courthouse in Detroit.

Not every innovation has been useful. Some unusual patents have been US 1744727, for a long speaking trumpet which projects over the hood to "facilitate traffic"; US 1865014, an attempt to push pedestrians out of the way of cars (even for 1930 it was a little archaic for the patent to refer to "horseless vehicles"); US 4315535 protects cars from flooding; while US 4805654 provides a massive sunshield over the roof. Some may be thought of as more useful, such as radial tires, the introduction of which drastically shrank manufacturing capacity (as they last several times longer than the old bias-ply tires). In addition every self-respecting car must be fitted with a cupholder. Suggestions for non-spill cupholders at the Halfbakery website are for ball bearings underneath the cup, which does not help when going round sharp bends, and inflatable bags. Surely easier is the concept patented in what may be the first cupholder patent for a vehicle, US 3842981. In 1973 Thomas Lambert of South Pasadena, CA, suggested a gimbal arrangement, as in boats, to ensure that the hot coffee does not spill. There have been at least a score since, not all of which are designed for cars. These include cupholders for supermarket carts (US 5838091) and strollers (US 5857601), possibly for the benefit of children sitting in them rather than those doing the pushing.

Not everyone loves cars, and some dream of replacing them with public transport, walking or cycling. One suggestion is the now famous Segway™ Human Transporter. Originally billed as Ginger, it was a mysterious project which its backers assured everyone would revolutionize city life. It has been called the most hyped high-technology concept since the Apple Mackintosh. On December 3, 2001 it was finally revealed that Dean Kamen, its inventor, had come up with a two-wheeled battery-driven vehicle which was controlled by gyroscopes (which ensure stability). Leaning slightly forward moves the scooter forward, leaning back reverses, and turns are made by twisting the handle. There are no brakes, as you merely stand erect. The batteries can be recharged overnight and the top speed is 12 miles an hour, which sounds hazardous on sidewalks, while its users would be vulnerable on the roads. There is a cluster of patent applications including US 6367817, filed in 2000. The idea is that pollution and congestion will be greatly reduced by everyone going around on the machines. Kamen said in *Time* that the Segway™ "will be to the car what the car was to the horse and buggy. Cars are great for going long distances. But it makes no sense at all for people in cities to use a 4,000-pound piece of metal."

Over 30 states have passed legislation allowing the Segway on sidewalks. These vary, with some restricting their speed, and Washington stating that they must give way to pedestrians. About half allow local municipalities to impose their own restrictions, and San Francisco for one seems antagonistic. The machines are for sale on Amazon's website for just under $5,000 and plenty of orders have been

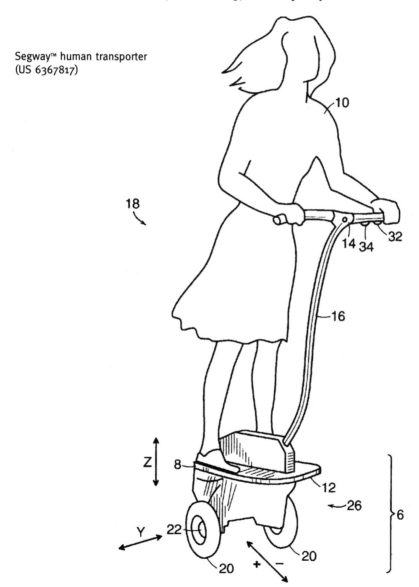

Segway™ human transporter
(US 6367817)

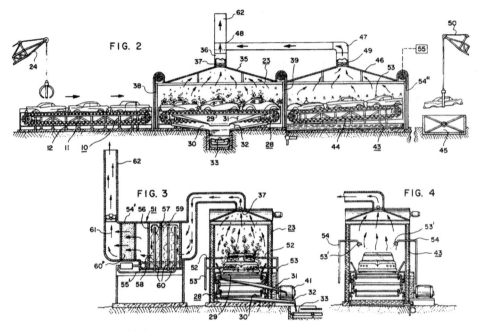

Car scraper (US 3613562)

taken despite the 10% deposit. Shipping from Deka's Manchester, NH, plant began in March 2003. It remains to be seen if this invention—without radio, heating, air conditioning, trunk and other accessories—will make a big impact, or if it will fade back into American legend.

All good things must eventually come to an end, and cars (at least in theory) finish their careers by being turned into scrap. An attractively illustrated patent, US 3613562 by the Garbalizer Corporation of America, filed in 1970 from Salt Lake City, is just one of countless examples of how to finish them off. Their "Vehicle body shell processing plant" was designed to minimize air pollution and was presumably inspired by the requirements of the Clean Air Act of 1970. Figure 2 shows the chain conveyor belt of vehicles moving along with Figure 3 being a cross-section of the first hall (the furnace) and Figure 4 of the second hall (the "quencher unit"), with doors at each end. The non-ferrous "metal droppings" were meant to go down (29) and (30) in the furnace to the pan (32) and hence by conveyor (33) on to further processing, having a lower melting temperature. The burnt-out steel carcasses (minus seats and the like) then moved on to the "quencher unit" where water poured on to them before going on to (45) where they were baled as compact pieces of steel for processing by scrap yards.

The customer is king

I T is extremely convenient to do all your shopping in a single place and then to take it away by car. Equally, part of the American Dream is success in selling a product or service, perhaps with a special gimmick or slogan. It is often said that the customer is king, although the customer might feel suspicious about this. It has been suggested that on the contrary people are not in charge—but their desires are. Manipulation, especially of children to take advantage of the pest factor, often seems to lie behind advertising to increasingly skeptical consumers, although the defence is that it is brands rather than the product itself that is being promoted. It is strange how often it is foods that are high in fat, salt or sugar that are advertised rather than vegetables or fruit.

Before something is sold the potential consumer has to know about the product or service. There are thousands of patents for advertising. There seems to have been a rule for many years that an actual name could not be given in patent drawings, as so many give John Doe as the name of the advertised product, even for, say, ladies' shoes (as in the 1928 "Advertising tower", US 1775267, which shows a young and sophisticated lady displaying the goods). Early advertising was not known for its subtlety, often using 100 words where 10 would have done. A good example of this is US D5068 by Joshua Brooks of Boston, "Design for an advertising print", which dates from 1871.

Despite its title the concept goes well beyond the appearance of the print and is really for a business idea. As the title indicates it was meant to be used both for advertising a particular business and for encouraging trade there, although the drawing offers a strange choice of subjects in each corner: a steam train, a stage coach, a fountain and an eagle. The "mercantile advertising bond" was meant to be handed over to customers who spent $5 and the bonds were to be redeemed at the Mercantile Advertising Bond Company to get a "valuable present" from the illustrated list shown. In other words it was an early (and inflexible) attempt at using trading stamps, which were not officially launched until 1891. Schuster's department store in Milwaukee, WI, then began issuing blue stamps to customers, who pasted them in booklets and redeemed them for cash. It is not known if Brook's idea were ever put into practice.

Some ideas are by contrast so subtle that they are not indexed under advertising, the most ingenious of these perhaps being by Stacy Carkhuff of Akron, OH. In 1908 he applied for what became US 1093310, the first non-skid tire. The invention quite simply consists of endlessly using the words "Firestone non-skid" as the tread. The wording in the patent's description merely talks of "engaging surfaces extended in lines or rows". Using oblique lines meant improved

Early trading stamp (US D5068)

road-holding ability (and was probably more eye-catching). The tread was also reinforced by using vulcanized rubber. The patent does not give Firestone as the assignee but it is known that Harvey Firestone suggested the idea.

Another, perhaps not so subtle, form of advertising occurred with the famous Buster Brown® trademark. Buster had been a character in an early newspaper cartoon strip by Richard Fenton Outcault from 1902, who dressed in Little Lord Fauntleroy clothes. John Bush, a sales executive with the Brown Shoe Company of St Louis saw the value of the Buster Brown name as a trademark for children's shoes (and also liked the coincidence in the "Brown"). He persuaded the company to purchase rights to the name from Outcault, and the brand was introduced to the public during the 1904 St. Louis World's Fair. Bush went on to become President

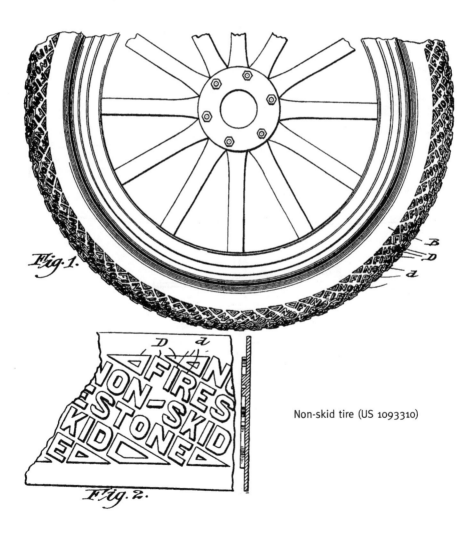

Non-skid tire (US 1093310)

and later Chairman of the company. Brown Shoe made marketing history when it sent on the road between 1904 and 1930 a series of midgets dressed as Buster and accompanied by Tige, his dog. They toured the country selling Buster Brown® shoes as they performed in theaters, department stores and shoe stores. The whole town would usually turn up to watch.

Bush went further. It is common now to incorporate a prominent trademark in clothing or shoes. An early example of this practice must have been US 1520224 by Bush himself, which dates from 1924. It is a "Means for identifying shoes" which uses "a type which has been sold on the market for many years" as its model, with a window inserted in the lining with the oval containing the trademark being secured at the back by discreet stitches (5). The general idea, of course, was what was being protected rather than the trademark itself. The cartoon is long forgot-

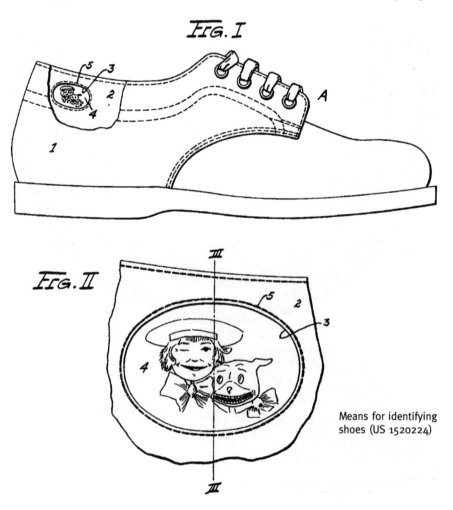

Fig. I

Fig. II

Means for identifying shoes (US 1520224)

ten but the trademark, and the concept of getting customers to pay to advertise the company, lives on.

Another form of advertising, long popular, was with matchbooks, where a necessity was given away as a form of cheap advertising. They date from 1891 with US 483165–66 by Joshua Pusey, a Lima, PA, patent attorney. He spent years defending his patents and then sold out to the Diamond Match Company for $4,000. They modified the idea before beginning manufacture, as Pusey had placed the striking surface inside the book, which meant that lighting one match also meant lighting the remaining 49! The earliest known commercial matchbook dates from 1895 and was distributed by the Mendelson Opera Company. The advertisement from the one surviving example was "A cyclone of fun—powerful caste (sic)—pretty girls—handsome ward-robe—get seats early". On the front was a photograph of Thomas Lowden, a trombonist. The opera had purchased several boxes of blank matchbooks from the Diamond Match Company, and the cast apparently pasted photographs and wrote slogans on the matchbooks.

Henry Traute, an enterprising salesman with the Diamond Match Company, secured some early lucrative orders from Pabst Blue Ribbon® beer, from Duke the tobacco industrialist and from William Wrigley, who ordered 1 billion matchbooks advertising Wrigley's® chewing gum. Traute was also responsible for the famous wording "Close cover before striking" on the flap. Matchbooks were sold for just one-fifth of a cent each, and the practice of giving them away when selling tobacco had become so widespread that the Office of Price Adminstration ruled in World War II that the presentation of matches with cigarettes had become so general, and so automatic, that the practice had to be regarded as mandatory.

This example, US 2184629, shown on page 162, was by Abraham Weiss of New York City and dates from 1937. Figure 1 shows the open book, Figure 2 the front of the book, Figure 3 the back and Figure 4 the open book with the cover lifted off. Prizes such as $5 or 5 gallons of gasoline were suggested (appropriately enough for the example, Socony, an oil company). Weiss did not say, however, what would be printed if a prize were not offered.

The act of movement in advertising as a means of attracting attention is much coveted. Skywriting is little used today but was very popular for decades as a dramatic advertising medium. The technique was first developed by John Savage, a British engineer and former RAF pilot. Letters 1 mile high and 1 mile wide were formed by specially built planes equipped with smoke-emitting apparatus. The engine's heat turned treated paraffin into white smoke, which was discharged under pressure. The "writing" is done at heights of 10,000 to 17,000 feet and is possible only in cloudless skies with, of course, little wind (unless it is a light wind blowing consistently). Contracts are commonly made for skywriting over a designated place such as a racetrack, fair, bathing beach, or carnival at a specified time. Savage moved to New York City as business opportunities were better than in Britain. His US 1489717 dates from 1923 and is one of his initial flurry of patents

Dec. 26, 1939.   A. WEISS   2,184,629

ADVERTISING MATCH BOOK

Filed Sept. 29, 1937

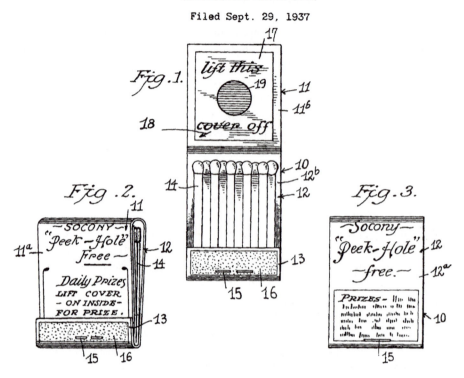

Fig.1.

Fig.2.

Fig.3.

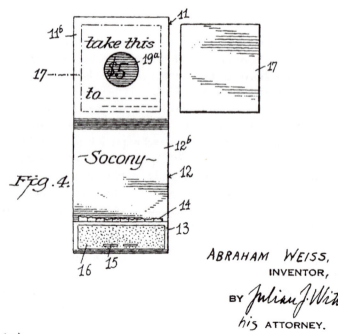

Fig.4.

ABRAHAM WEISS,
INVENTOR,

BY *Julian J. Wittel,*

his ATTORNEY.

Matchbook (US 2184629)

covering the idea. Here, having worked out how to produce the smoke, his concern was with how to maneuver the airplane to produce legible figures, in this case, appropriately enough for smoke, the words "Lucky Strike". Savage used an adapted SE5a, the best British fighter plane in World War I, for his skywriting. When his team visited New York City 'Hello USA' was written in the sky and hung in the air for 10 minutes, bringing the city to a halt. Blimps later became much more common for aerial advertising as they were less susceptible to the weather. They also have the bonus of being able to provide television coverage of major events.

Several hundred patents cover the idea of moving vehicles carrying advertising. These vary enormously: some show the advertising placed on the door, or above the doors, or on the license plates, for example. A few were more extravagant, and envisaged animation. An example is US 2091670, filed in 1937 by Denison Budd of North Tarrytown, NY.

Budd explained that delivery trucks have a large amount of space on the outside bodywork which is not being used for advertising the product carried inside. A small auxiliary motor provided the power for the animation as Budd dismissed the idea of using power directly from the engine. That would have meant the animation stopping each time the truck halted, and when the truck was moving quickly or slowly, "the actuation of the moving parts was too fast or too slow to be

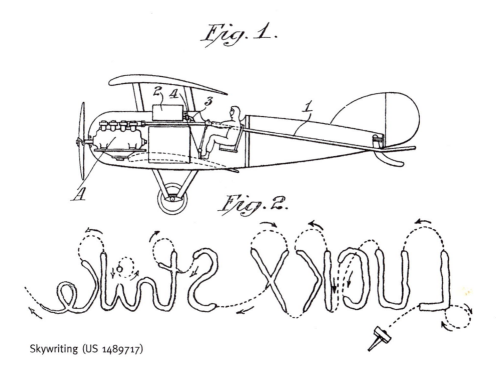

Skywriting (US 1489717)

effective". He suggested that trucks could advertise other products and not necessarily those of their own company, although not all products lend themselves to animating as the act of eating does. Budd claimed to be the first to think of such a concept that would work without taking up space within the truck.

The entire display was mounted on a frame secured to the side of the truck. The motor was placed between the frame and the side. A simple clockwork mechanism enabled the parts to move, in this particular example the head and the left arm. Budd explained the effect in case it was not clear. The drawings showed that "the gentleman at the table is consuming the pancakes with great expedition and gusto inasmuch as he appears to be transferring them from the plate to his interior most rapidly and enjoyably".

Another means of using movement to attract attention is illustrated by US 2935806 by Manly Young, Sr. of Fort Mill, SC, who filed for his patent in 1958. It

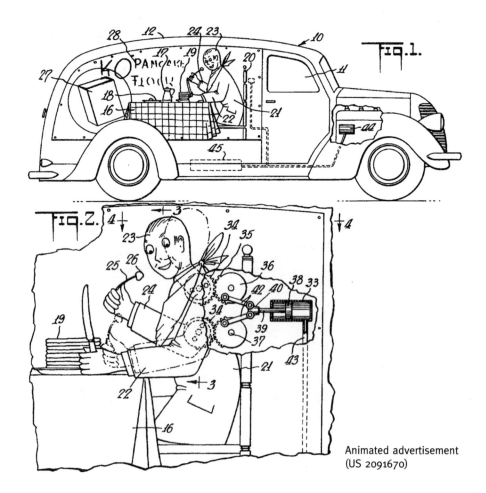

Animated advertisement
(US 2091670)

Poster advertising (US 2935806)

is now normal to see advertisements on billboards changing at intervals. This can be done electronically, or by numerous cubes making up the picture revolving to create a fresh image, or by the advertisement sliding out of sight while a new one replaces it. This both attracts attention and of course boosts the money that can be charged as there is more than one customer. This must be an early example, but the appreciative looks from those watching are unlikely to have been experienced in reality.

Young explains that moveable bands operated by rollers would at intervals change the advertisements, with two placed side by side for economy. He had thought through the problems: everything was encased in a housing which was "waterproof, dustproof and substantially ratproof and bugproof". To prevent pests, presumably, it was recommended that a sheet metal casing be used for the whole apparatus. A marquee was provided above as shelter from glare and rain and could house neon lights, and the glass frontage was tilted outwards to prevent dust or moisture building up.

Advertising has of course changed dramatically with specialist cable channels and a vast amount of intrusive advertising on the internet in the form of "pop-up" advertisements and emails selling anything the viewer imagined, and much more besides. If this advertising did not exist charges would have to be paid by users as somehow the cost of the internet has to be paid for. By contrast US 3603592, a game in which you have to recall the advertising you have seen on television, seems quite innocent and acceptable. Joe Bury of Wyland, WY, had filed in 1970 for his "Apparatus for playing a game using the perception of television commercials".

Sometimes a bit of tinkering is needed before a slogan reaches iconic status. The famous "When it rains, it pours" slogan for table salt made by what is now Morton International, with a girl holding an umbrella in one arm and an emptying package in the other, has been on salt containers since it was registered as a trademark in 1915. It originated in the company's first advertising campaign (with N.W. Ayer & Company, also of Chicago) after they invented the first freely flowing salt following 4 years of research. The solution was to add some magnesium carbonate as an absorbing element, which meant that the salt would not get lumpy in damp weather. Previously people had had to smash up what they needed with anything to hand. A metal pouring spout was also added to the new blue cylinder. All this cost money, and as the product could no longer be sold for a nickel a campaign was needed to encourage sales at a higher price. The proposed picture suggested the virtue of the product, even if many nowadays do not realize the significance. Although Sterling Morton loved the drawing, he was less keen on the proposed slogan to go with it: "Even in rainy weather, it flows freely" did not have quite the right touch. Later suggestions included "Flows Freely", "Runs Freely", "Pours" and finally, an old proverb, "It never rains, but it pours". The wording seemed too negative, and putting a positive spin on it resulted in the famous slogan. The campaign first ran in *Good Housekeeping* and was a success. The girl's dress and hairstyle has had to be updated at intervals to keep up with fashion—in 1933, 1941, 1956 and 1968. She now strides out, and has also switched from being a blonde to a brunette and back to a blonde. The original trademark is shown.

Original When it rains, it pours trademark

The girl herself came out of the artist's imagination. Many trademarks in the old Patent Office gazettes consist of sketches of solemn looking people advertising long forgotten products, often medicines. Typical wording is "said trade-mark consists of the portrait of Dr. L. Montgomery, deceased, and the signature of said Dr. Montgomery", in an advertised trademark in the gazette for January 12, 1915 for a remedy for 10 enumerated ailments (including "nervousness"). No model was used either for

Betty Crocker®, who originated in 1921 as an imaginary respondent to questions sent in to the Gold Medal Flour Company. She owed her surname to a much-liked former company director, while her first name was chosen because it just sounded nice and friendly. Like the Morton girl her appearance has changed with the times, with seven versions in her case. She is meant to be a pleasant woman who will not make others feel inadequate when dealing with household tasks.

On the contrary, the famous Gerber baby, with her open mouth, was modelled on the baby of a neighbor of Bostonian artist Dorothy Hope Smith. She entered the charcoal sketch in response to a competition when the Fremont Canning Company was planning a launch of "canned vegetables" baby food. Although she said that it was unfinished, the company liked it so much that they used it from 1932 and registered the image as a trademark with the note "the picture of the child is merely fanciful". Aunt Jemima® too is based on a real person. From 1889 her image was based on Nancy Green, who was born into slavery on a Kentucky plantation, and who spent over 30 years travelling around the country promoting the virtues of the Pearl Milling Company's self-raising flour (Quaker Oats took over the company in 1925). Her career was only ended when she was killed in a traffic accident in 1923, and a new model, Anna Robinson, was depicted on the packaging for several decades. The name itself was taken from a vaudeville song of the time. Another non-threatening African American icon is Uncle Ben®. His appearance was based on Frank Brown, the maître d' of a Houston restaurant, but the name is thought to be based on a local rice farmer who was famed for the quality of his product. The name was first used in 1937. As for the Campbell Soup Kids, with their delighted eyes and ruddy cheeks, they were drawn in 1904 by Philadelphia artist Grace Wiederseim, who claimed to have modelled them on herself. They have grown both taller and slimmer since their first manifestations as streetcar advertisements.

Edison once said, "Anything that won't sell, I don't want to invent". Relatively few do invent, but quite a few proceed to sell to the many who were clearly born to shop. In 2000 about $2.2 trillion was spent in retailing, which was 41% up in real terms from 1996. Clearly more and more is being spent but to the regret of many the personal touch of the old corner store, as depicted in many a Norman Rockwell print, has surely been lost forever, at least in the more prosperous parts of the country. Ironically, with more efficiency the number of people involved has grown enormously. In 1870 for every 1,000 workers producing goods there were 88 involved in distribution (which includes deliveries). By 1980, this had changed to 675 being in distribution.

Despite trends towards "e-tailing", shopping malls remain a fundamental part of American life. The concept of developing a shopping district away from downtown is generally attributed to Jesse Nichols of Kansas City, MO. His Country Club Plaza, which opened in 1922, was built as the business district for a large-scale residential development. The architecture was similar in style, it had paved and lighted parking lots, and it was managed and operated as a single unit. But many consider Highland Park Shopping Village in Dallas, developed by Hugh Prather in 1931, to be the first planned shopping center. Like Country Club Plaza, its stores were built with a unified image and managed by a single owner, but it occupied a single site and was not divided by public streets. The stores also faced inward, away from the streets, which was a revolutionary concept. These early attempts would not have worked without a high density of car ownership, which explains why shopping centers first started in America and not in Europe.

The actual display of the goods is vital to keep sales high. The world's leading greeting card company is Hallmark, named for Joyce Hall, its founder, and based in Kansas City, MO. The company's name was originally Hall Brothers and they first began using Hallmark as a trademark in 1925. It was a pun as a "hallmark" was originally used by London goldsmiths to guarantee the purity of the product. They were the first company to use a brand to sell greeting cards. The company was innovative from early on, in 1932 licensing the rights to use Walt Disney characters, which was the first such venture for both companies. Cards, though, were still kept in drawers which discouraged impulse buying. Often the store assistant would be asked to suggest a nice card. The idea of display was revolutionized by Joyce Hall's own US 2067051 in 1935, which although revolutionary for its time now looks very ordinary and dull.

For the first time people could browse through the cards themselves while their attractive "faces" were easy to see. The display rack is not set at a single angle but rather at a varying slope, which makes the display easier to view. Glass panels hold the cards up in each row. A more sophisticated and modern-looking adaptation was US 2810483 by the company's Willard Kallhorn which dates from 1954. It specifically refers to the older patent and points out that a disadvantage was the very equality of display: every card was treated the same and none were highlighted. The basic idea was kept but adjustable racking was used to present entire cards (presumably those with higher profit margins, or for current festivals) in front of the rest so that the entire card could be seen rather than the top two thirds.

The company has had to continue to innovate. They have introduced machines in their stores where customers can personalize cards and print them off with patents such as US 5036472 which dates from 1988. A major threat is the trend towards using free websites to send one of a wide selection of cards electronically with, again, personalized messages and often very attractive animation. In March 2002 Hallmark announced a deal with Tumbleweed Communications Corporation of Redwood City, CA. Hallmark.com will license Tumbleweed's technology for delivery of greeting cards over the internet. The patents include US 6192407 for "Private, trackable URLs for directed document delivery" and US 5790790 for the "Electronic document delivery system in which notification of said electronic document is sent to a recipient thereof". It will be a challenge to sell something which rivals are giving away.

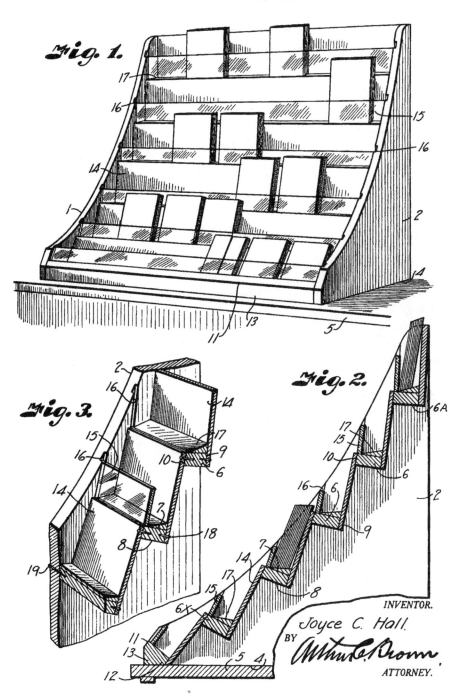

Fig. 1.

Fig. 3.

Fig. 2.

INVENTOR.

Joyce C. Hall

BY

ATTORNEY.

Card display apparatus (US 2067051)

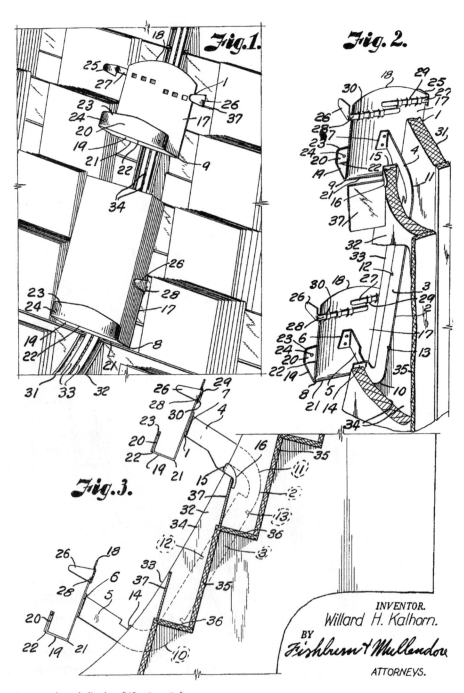

Fig.1.

Fig.2.

Fig.3.

INVENTOR.
Willard H. Kalhorn.

BY Fishburn + Mullendore

ATTORNEYS.

Improved card display (US 2810483)

Another innovation in shopping was invented by Sylvan Goldman of Oklahoma City. He and his brother moved into the wholesale retailing of produce, only to be wiped out when oil prices crashed in 1921. They went to California to learn about retailing techniques and then returned, starting again with self-service stores and woven baskets for the customers to carry their groceries. They prospered and sold out to Safeway, only for the value of their Safeway stock to plummet with the Depression. They started again, remembering that "the wonderful thing about food is that everyone uses it—and uses it only once", and prospered again.

One night Goldman sat in his office wondering how he could entice his customers to buy more at his Humpty Dumpty store. He had noticed that customers would head for the checkout as soon as their baskets were filled. He stared at a wooden folding chair and imagined a basket on the seat, and wheels on the legs. Together with a mechanic, Fred Young, he worked out the details of an X-shaped structure with wheels designed to fold sideways, with space for a removable basket at the top and for another on the bottom. At the top, the "back" of the chair-like structure functioned as a U-shaped handle. His US 2196914, filed in 1938, was a "novel rollable market basket carriage of a lightweight portable type expressly but not necessarily adapted for convenient usage by shopping customers in grocery stores". After individually folding away each cart the baskets could be stacked up. This was important in saving space after closing time.

Goldman set up the Folding Carrier Basket Company to manufacture his new carts. On the first day they were available he waited to see them being used. At first his customers were skeptical: they looked at the carts but did not want to try them. When he asked why they did not want to use them reactions varied. Some men thought their use was effeminate, while young women tended to think that they were unstylish. One asked if he thought she was not strong enough to carry a basket. Goldman resorted to hiring people to push them around to demonstrate how they should be used, and employing "greeters" to show anyone coming in how to use them. Once the customers realized how much more they could carry both the idea and sales took off. By 1940 the concept was well established, and there was a waiting list of 7 years for supermarkets to purchase carts. There was one drawback: the aisles and checkouts often had to be modified to allow them to pass.

The idea did need modifications. The new carts were still relatively small, and customers had to bend down to put purchases in the bottom basket. Their design also meant that time was wasted folding and stacking at the beginning and end of the day. Goldman tinkered some more, and finally filed in 1949 for US 2556532, a "Store service carriage", the modern shopping cart with a large wire tub and a rack beneath for bulky purchases. He also obtained US 2508670, a removable baby seat for the same (as he had noticed problems for new mothers). The improved design meant that they could be easily "nested" by running them into each other. If you take away the wire basket, of course, you have something very like the typical cart for moving suitcases around at airports. The design has been little changed since.

One early improvement was fixing the rear wheels so that only the front wheels can swivel. This means that the cart moves smoothly round corners. This is unlike the British model, where all four wheels are in action, making cornering a desperate struggle against forward momentum. In 1954 a modification allowed the name of the store to be given in large letters on the handle. Chrome finishes meant that the carts could withstand rain if kept outside. Small wire baskets within the main baskets for more delicate items are also common. Goldman's company still thrives, now under the name of Unarco. Goldman himself died in 1984, worth $400 million.

Supermarkets clearly want to make the experience of paying for, and packing, goods, as painless as possible. US 4953664 is an unusual manifestation of this. It was filed in 1989 by Sonoco Products Company of Hartsville, SC. The title was "Ergonomically designed check-out counter system for supermarket and merchandising industries". The patent discusses in some detail the shift from mechanical cash registers to "integrated point of sales terminals" in the 1970s which resulted in increased work rates and reduced labor requirements. A problem from the shift was that the cashiers could suffer from health problems, leading to lawsuits. The main ailments were strains and sprains in the lower back and in the arms. The patent proposed a solution.

Most check-outs in North America are designed so that the sales assistants work while standing up (in Europe they tend to be sitting). The patent suggests that they can sit if they wish, with the height of the work surfaces being adjustable to allow for individual height and reach by activating a motor which moves the floor underneath the seat with a mechanical jack. The goods would be placed in the bags by the cashier ("direct bagging") rather than by a bagger, so that each item was only handled once. When each bag is full the cashier activates the conveyor belt. As it moves off a deposit of glue on its back tugs the bag behind it so that the second bag opens up on the arms (73). Bulky items would understandably be left on the counter (72) for the purchaser to put away. The keyboard could be moved to allow for left-handed sales assistants. The bags were presumably meant to be of paper: in Europe groceries are much more likely to be packed in plastic bags.

We take the price tagging of goods for granted. In the old days you bargained, and fixed prices had the advantage that children sent down to the store were not going to get cheated. Just the same it was tiresome writing the price on everything and attaching it to something. Monarch Marking Systems is the world's leading merchandise tagging company, and it began with Frederick Koehle of Dayton, OH, who in 1890 filed for US 457783. The way this device originally worked may seem little better.

Clearly, it was best used for items like textiles where pins could conveniently be fastened, and the patent indeed says so. A piece of heavy-duty paper was folded over, and then each end was folded over to reinforce them. A piece of metal was held in place within one of these folded ends. The portion of the flap over the

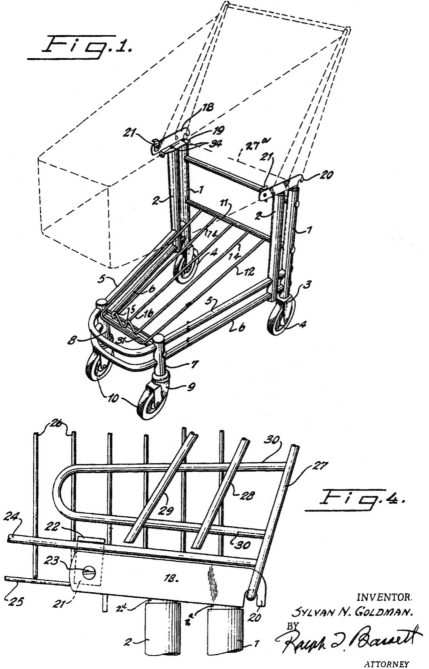

Fig.1.

Fig.4.

INVENTOR.

SYLVAN N. GOLDMAN.

BY

Ralph J. Barrett

ATTORNEY

Shopping cart (US 2556532)

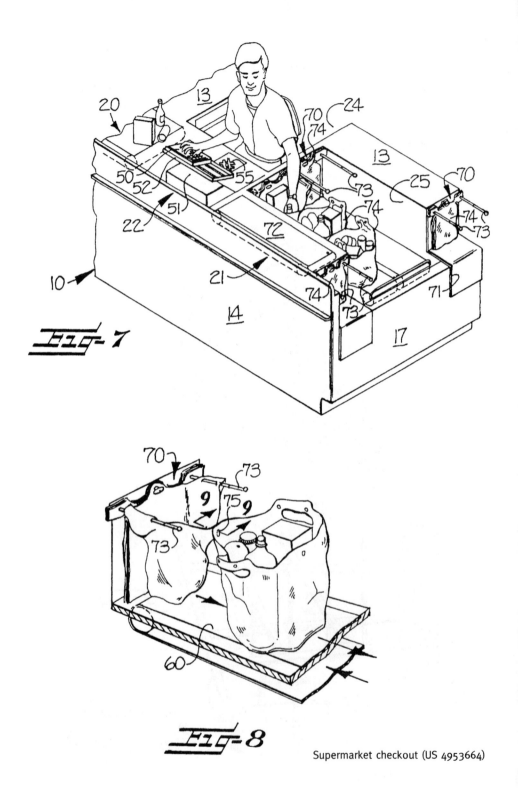

FIG-7

FIG-8

Supermarket checkout (US 4953664)

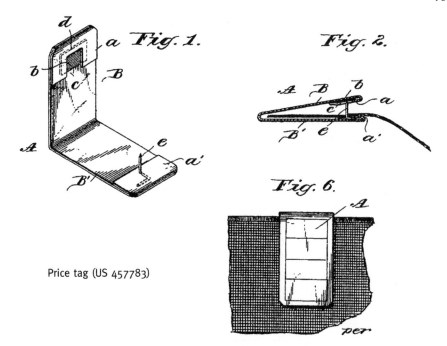

Price tag (US 457783)

metal (B) was cut out and a pin (e) was inserted at the other end, as shown in Figure 1. The fabric was then placed between the two ends and fastened together as in Figure 2. With the two sides of the tag showing, as in Figure 6, they could be written on with the price or other information. It was primitive but it was also a distinct improvement on previous efforts. Kohnle remarked that previous inventions obscured part of the label with the "attaching devices". He also suggested a variant where a staple on one flap received the pin, which is much closer to modern ideas. The trouble is that this patent is credited with automatically printing the price as well and that is not covered by it.

By 1904 the company had introduced an electric tagging machine working along the same principles of the pin. Constant innovation was needed to keep up with retailers' expectations and to ward off any competition. Nowadays we are used to seeing staff scanning bar codes for information such as sell-by dates and punching labels with handheld machines, with the Monarch Pathfinder in 1984 being the first such device.

An example of taking pains to sort out problems is the same company's introduction of pressure-sensitive labels in the late 1940s. They were meant to replace the old gummed labels, which were useful for goods that could not take pins. Their US 2636297 for pressure-sensitive labels, applied for in 1950 by Floyd Johnson, indicates the efforts taken to ensure a usable product in a field nearly everyone takes for granted. It was worth it to secure such a large and lucrative market. The patent was an improvement on US 2095437 by Louis Fox of Los Angeles, which dated from 1936. This was a method of making a strip of labels, with lines of weakening between each label, the weakening being aided by little rectangular apertures at each side. The strip was coated with adhesive which was masked by backing paper. The problem was that the labels were often torn when removing the backing. This was tracked to the fact that the little apertures were made in the strips *after* the backing was in place, so that the edges of the labels often got tangled with the adhesive, and the weakest element—the label—gave way when the backing was pulled at. In addition, the apertures by cutting into the labels at right angles weakened the strip so that peeling the backing off often damaged the labels.

An attempted solution had been to place the adhesive in two bands which crossed the apertures and ran the length of the strip. This meant that the adhesive was not present at the lines of weakness. Customers then complained that the labels tended to curl up and come away, but the new version was still used extensively, "the tendency for the labels to come off the articles being less of an evil than the difficulty of peeling encountered when a solid coating of adhesive was used". The 1950 patent solved these problems by using V-shaped apertures and a diamond-shaped hole in the middle of each line of weakness, all of which encouraged easy peeling.

Internet shopping is a pain-free method of shopping for many, and in an attempt to make it seem as close as possible to normal shopping there are often icons of shopping baskets or carts, and you are asked if you wish to add something to the shopping cart, or to proceed to checkout. Jeff Bezos' Amazon was the first big internet retailer. In 1994 he was a computer scientist working on Wall Street when, looking for business opportunities on the internet, he decided to sell a product which was already listed electronically: books. It is an area where people usually know what they want (or sample pages can be displayed), and where items do not get rancid or otherwise easily damaged in the post. Music is just as good if the user can hear sample tracks. Clothes or food can be more awkward, as people usually want to try clothes on or to taste the food.

Amazon's famous "one-click purchasing" patent was filed in 1997 by Bezos and three others from their Seattle headquarters. US 5960411, a "Method and system for placing a purchase order via a communications network", explains in detail how John Doe can purchase his desired product over the web by using the "shopping cart" model of a virtual supermarket—you identify yourself, pick up your goods, and you pay. What was new was that by placing a "cookie" at the user's PC the system could subsequently identify both billing and delivery information. The next time the customer places an order this is identified and the order is processed quickly and efficiently (and, some might argue, it also makes the users' computers more vulnerable). This is very attractive as besides saving the time of both customer and retailer, it prevents many lost sales. It has been estimated that half of all "shopping carts" are abandoned before completion of orders, often through frustration at using the internet.

Many are furious at the idea that someone should expect royalties for "an important and obvious idea for E-commerce" and urge boycotts of Amazon. BarnesandNoble.com was taken to court during the busy Christmas 1999 buying season by Amazon for using a similar system where by a single click pre-registered customers could use an "Express Lane". The legal arguments were not about whether the idea was thought to be unpatentable, but rather whether prior art showed that the idea was not new. A settlement in March 2002 ended the dispute, but as it was undisclosed it leaves the patent still hanging over other web retailers. It looks as if it included adding a second click stage to the procedure.

Nobody really knows how much money is spent in hyperspace. Every e-tailer seems to give contradictory figures. It has been suggested that the market was £2.4 billion in 1997, rising many fold over the next few years. Few yet seem to be making a profit, but it seems to be the place for retailers to be. It remains to be seen if the internet is the future of retailing, with virtual shopping malls scattered across hyperspace.

Beauty is skin deep

THE American Dream is interwoven with clothing and fashion accessories, although it is odd how so often the less material there is the more it costs. Manufacturers are as keen to sell as the public are to buy. Of course, it would save a lot of money for the manufacturers if they did not have to produce so many sizes in the different styles available. Just the same, customers often grumble that although they like the style none of the clothes fit. It is ironic that in the bad old days customized clothing was precisely what you got. Factory-made clothing only came about because of sewing machines. Some years later there was also the stimulus of needing to produce vast amounts of standardized clothing in the Civil War for the Union army. This need was partly provoked by the confusion resulting from many Union regiments wearing Confederate grey and not Union blue. Manufacturers were given the most common proportions, and they began to produce regular sizes which would accommodate most men. By 1880 half of all clothing was ready to wear. These "hand-me-downs" were at first despised, but were soon recognized as good quality.

Now work is going on in the field of customized clothing. The problem is capturing a person's body dimensions so that a piece of clothing can be made to give a good fit without (says a Levi Strauss patent application) "undue expense for the manufacturer, retailer or consumer". The fact that clothing when worn is in three dimensions makes them complex to make, as changes in one dimension (such as the waist) may require changes in another dimension to ensure a good fit. The increments between sizes are typically large (the "2" in the difference between size 30 and 32 pants for example is quite large) yet the cost of providing many more as stock at the retailer would be prohibitive.

The same Levi Strauss application (US 2002/123821) soberly points out that customizing clothing is traditionally done by one of four methods: a tape measure; placing the garment on the person and making tailoring adjustments; using an adjustable garment; or a mechanical, optical or video device to capture body dimensions so that a unique pattern is created from which the garment will be constructed. The last seems to be the most promising method for further research. An example of how this might work is US 6415199, which was filed in 1999 by E–Z Max Apparel Systems of Narberth, PA.

While discussing the same problems the patent adds that personal fittings are intrusive and discourage many buyers. The invention is for a stretch bodysuit with lines on it which allow the dimensions to be captured by optical scanning. A computer program then uses the dimensions to generate a pattern for the desired garment. The exact fitting is now known to the retailer, which is valuable information—although overindulgence or other factors may make the data useless. Perhaps in the future everyone will have such a stretchsuit so that we can communicate direct with the manufacturer via the internet.

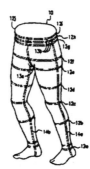

Enabling customised clothing (US 6415199)

Alternatively, the idea of mixing clothes to make new combinations may appeal. US 6161223, "Pants separable at crotch for style mixing" by Allison Andrews of Longwood, FL, in 1999 claimed to be the first which allowed separate halves of a pair of pants to be unzipped so that different colors and designs could be mixed. The idea only worked, of course, if her special products were bought. Another way of standing out from the crowd would be wearing clothes that change color according to the temperature of the wearer. In 1986 Donald Spector of Union City, NJ, filed for US 4642250 where the fabric of the garment was tailored to make direct contact with the skin of the wearer at

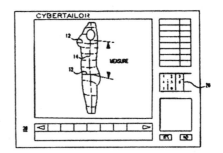

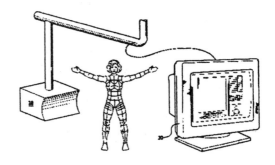

different "body sites". Integrated with the fabric were cholesteric liquid crystals which reacted to heat and changed color. No doubt the sites were chosen to minimize possible embarrassment. A simpler way to advertise yourself would be by wearing a "skin stencil" bandanna which spelt out a message after a lengthy sunbathing session, as in US 5652959 by Michael Proctor of Midway, UT, in 1996.

Styles can also be easily changed by turning some garments inside out (at least those which are designed for that). This can have a practical advantage if that changes the properties of the clothes, just as some comforters change their degree of warmth if turned over on the bed. Clothes that adapted to changing temperature and humidity would be very useful. One suggestion that only partly helps was made by Katherine Ball of Dallas in 2001 with her US 6480112. The "Clothing management system" meant that the clothes had to be arranged in a set order while a sensor pointed to the appropriate clothing. It is unclear what was supposed to happen if the weather were going to change, or if it were going to change in the area or indoor environment where you were going. Another idea now available uses nanotechnology. Costing $35, Lee® High Performance Khakis are coated with millions of tiny fibers, each one hundred-thousandth of an inch long, which are coated with water-repellents (and which resists wrinkles). This Nano-care® technology comes from Nano-Tex of Burlington, VT, which have a number of patents in the field, and they claim that it will retain its qualities for 50 washings. A skeptical TV station (WTIC-TV) tried spilling coffee and red wine on the pants with excellent results as the liquid rolled off, although pizza and ketchup stains did need some extra help.

Moving through different kinds of clothing from the head down, baseball caps used traditionally to be worn only by catchers and co-pilots of freight planes but are now an everyday adornment for many. They originated in 1869 with the Cincinnati Red Stockings. Some hats are specially equipped or decorated, such as US 2136925, a combined beach hat and fan; US 4551857, a solar powered cap to cool you down; US 4858627, a smoker's hat to prevent annoying others, complete with exhaust system; US 5457821, a hat simulating a fried egg; US 5542129, a convertible hat and catching glove; US D321274, a hat incorporating skyscrapers; US D336357, a hat emulating a football gridiron; and US 6101747, a ski hat incorporating a ski pole. The interest in sport is obvious. The urge to patent so much in hats is probably due to the perceived need to combine different activities in such an obvious and conveniently placed item of clothing. Hence US 5996127, which is for a hat, also acts as a feeding perch for birds, enabling close observation.

The splendid hat shown by John Geddie of Charlotte, MI, in US 4681244, from 1986, a "Portable bar", was meant to be "both useful and entertaining". The great number of tubes and valves is for mixed drinks, which could be conveyed directly to the mouth by tube (58), although it is unclear how the mixing could be done while the hat is being worn. A later, rival idea in the same field was US 5966743 where the wearer had a beer barrel as headgear.

Nearly all inventions in clothes are, like these, unsuccessful. One that continues to be used was invented by 13-year old Chester Greenwood of Farmington, ME. His ears used to freeze when he went skating, so he took loops of baling wire and with the help of his grandmother padded the ends with beaver fur. His friends laughed to see him with these strange "earmuffs", but they soon realized that he was able to skate for much longer, so they asked him to make some for them. In 1877, aged 17, he applied for US 188292.

Of all the devices used in hairstyling the meek bobby pin must seem to be the simplest. Over a hundred variations have been patented. Controversy rages on the web about who invented bobby pins, with dates ranging from 1916 to 1928, and

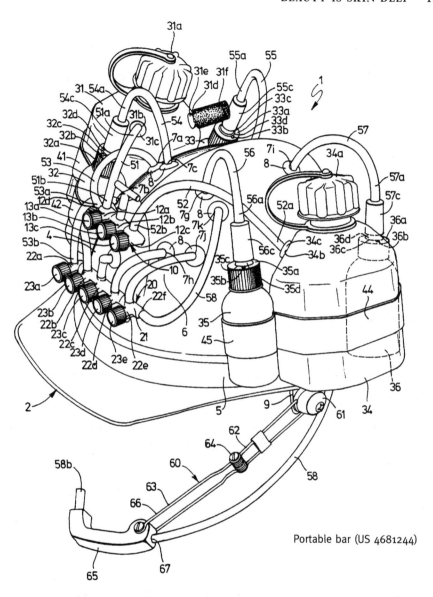

Portable bar (US 4681244)

Omaha, Danville, PA, and New Zealand all being acclaimed as the place where the idea first saw the light of day. Their popularity does seem to date from shorter, "bobbed" hair for women in the 1920s. The earliest patent to use the distinctive crinkled bobby pin look appears to be US 1233195 by Samuel Creech and Perry Bland of Sullivan, IL, in 1917. They simply called it a "hair pin". As for ensuring the correct hair style at the hairdressers, in 1941 Lincoln Newspaper Features of New York City filed for US 2309390, a "Photographic method of matching and

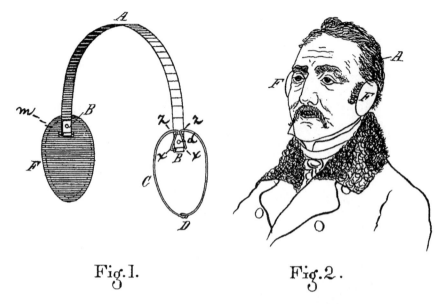

Fig. 1.   Fig. 2.

Earmuffs (US 188292)

selecting coiffures or the like". It enabled a composite photograph to be produced of the client with any of many hairstyles. The basic idea sounds very useful.

The first "globe" hair drier is thought to date back to Louis Ruffio of New York City who filed in 1923 for a personal machine type, later amended and printed as US Re 17447. The more economical idea of several women sitting in line, each with a hood over them, took a little longer, with an early and delightful example being US 1998924 by Frederick Crook and Eugene Caster of Kansas City, MO, in 1931. Evidently the inventors had not thought of the clientele reading magazines while they were waiting.

The large contraption on the right is a gas burner, complete with flue, which sent the hot air to each hood. There a baffle ensured that the hot air did not directly blast the head.

Turning to sunglasses, one claim is that they date from as far back as 1885, though in a primitive form, being made from tinted window glass in Philadelphia. Another is that Chinese judges in the thirteenth century wore them to protect their eyes from the sun; another that James Ayscough in Britain made them with green- or blue-tinted lenses in the 18th century to correct for vision. They did not become popular until the 1930s—almost inevitably, as seen on relaxing movie stars. Some can look a little strange, such as US 4909620 in the shape of moose antlers, and US 4919529 for rear-looking sunglasses. The idea was not new, as US 1691789 from 1926 was a much earlier attempt to help drivers see behind them. Several, such as

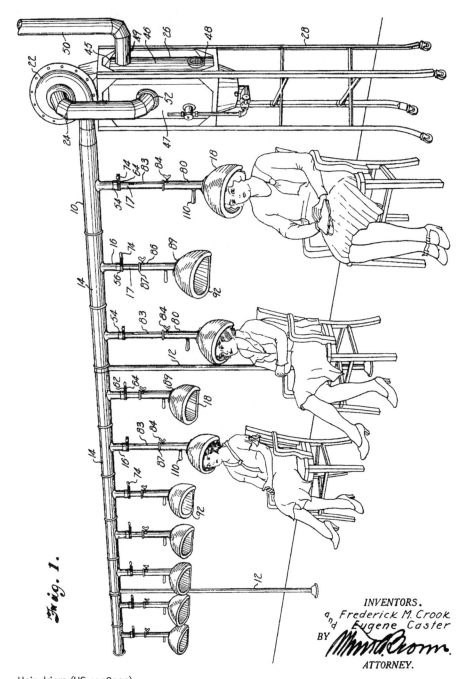

Fig. 1.

INVENTORS.
Frederick M. Crook
and Eugene Caster

BY *Wm. A. Brown.*

ATTORNEY.

Hair driers (US 1998924)

US 1806336 by Klara Karwowska of Calgary, are for the famous windshield wipers for spectacles. This 1956 attempt was called an "Eyeglass wiper".

Coming to coats, we take hangers for granted (unless clothes slip off or get creased on them). Like everything else they had to be invented. The idea dates back to 1869 at least with O.A. North of New Britain, CT, with US 85756. This has the classic triangular shape, but North wanted it to be secured to a plate on the wall. The imaginative leap to a portable device which was easier to maneuver clothes onto is credited to Albert Parkhouse.

Parkhouse was born in Ontario and moved as a boy to Michigan. There he got a job with the Timberlake Wire and Novelty Company, located in Jackson. The family's story is that on a cold day in 1903 Parkhouse returned from lunch and found that there were no coat hooks free on which to hang his coat. This often happened, as there were not enough hooks for all the employees. Parkhouse disliked the idea of placing his coat on the back of a chair where it would get wrinkled. There was always plenty of wire around, so he took a piece and twisted it so that it would fit inside the shoulders of his coat. Then he bent another wire into the shape of a question mark and secured it by coiling it round the top so that it formed a hook. There was still the problem of where he would hang it from, of course. His original idea was suitable only for hanging a coat, as he envisaged the bottom of the wire coming up to join the hook. He continued to refine the idea over the next few weeks and soon the other employees started using copies provided by Albert. The drawing shows a rather more elaborate version than is normally considered necessary, being engineered to deal with heavy coats. The helical spirals at each end are in the form of loops to help the coats preserve their shape.

The Timberlake company, like many companies at the time (and indeed today) were accustomed to having the inventions of their employees assigned to themselves so that they could profit from them. In this case they went one stage further. When in 1904 US 822981, "Garment hanger", was applied for it gave the inventor as Charles Patterson, who was actually the company's lawyer. Clause 8 of the Constitution is for "securing for limited Times to Authors and Inventors the exclusive Right" to their inventions, so this was unconstitutional, let alone illegal. Parkhouse is said to have been "a little bitter" about Patterson being named as the inventor, with the truth unacknowledged except to a few at his workplace. He soon departed with his family to Los Angeles, where he founded his own, modestly successful wire company. Parkhouse had even thought of the problem of pants slipping off. (15) on the drawing is a clamp for holding pants. Nowadays plastic coat hangers often have ribs above the bottom portion so that the pants can be firmly gripped. This can cause creases, and an effort to avoid them was the popular variant where a long cardboard tube encloses the bottom wire of the hanger. This dates back to 1931 with US 2023443 by Elmer Rogers of Detroit. There have been thousands of garment hanger patents since, all looking for simple and economical solutions to the problem.

Coat hanger (US 822981)

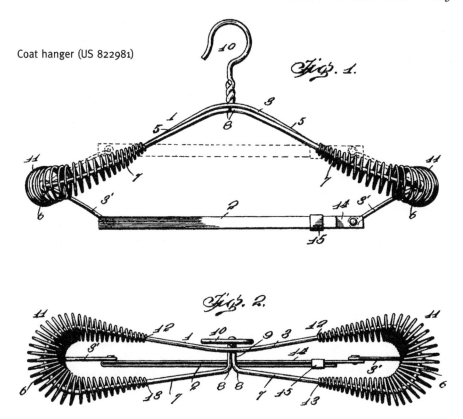

Suspenders seem always to have been around, if Hollywood movies can be trusted. One theory for their invention is by retired farmer Peter Leith of the Orkneys, the islands in the north of Scotland, who found recordings from the 1980s of two 90-year-old men describing an eccentric local named Davie Taylor. They told how Mr Taylor invented "a bib and brace sort of clasp for hooking his breeks up". Mr. Taylor was an "unemployed draper" in the 1891 census which also revealed an apprentice draper called Andrew Thomson, 17, who lived in nearby Stromness. Sure enough, in 1896 a patent was filed from San Francisco for a "Clasp for garment-supporters" by Thomson, a "citizen of Scotland", with his US 567421. Many Scots claim that their country invented everything and in this case, among the many earlier patents in the field, the credit for the first patent for modern suspenders is usually given to David Roth of Cleveland, who had already applied 2 years earlier for his US 527887, a "Garment-clasp". The general idea of attaching or closing garments has been the subject of furious bouts of patenting over the years, with buttons used for a long time. Press-stud buttons had already arrived, having been patented by Pierre-Albert Raymond of France in 1886 for the use of the glove trade. Nowadays Velcro® is often used more than zippers, being handier when using gloved hands, as well as for small children and for those with arthritic hands. Zippers on the other hand are better for keeping out the draught but can get stuck. Perhaps someone will combine their virtues in a new product.

It is difficult to credit the vast amount of attention paid by pre World War I inventors to corsets. Women were thought fragile and apt to faint, but this was no wonder when they were so short of breath. An experiment with volunteers wearing 1870s corsets verified the horrors of wearing one, with reduced lung capacity being clear. There was also the real fear of being punctured by the metal used in the corsets. The ideal waist size in the 19th century was 18 to 20 inches, and it was said that you had to get married at an earlier age than your waist measurement. Presumably it was men who were responsible for the pressure towards the look, and changing fashion as well as protests and the invention of the brassiere led to its slow demise. Nevertheless for many years there were some 25 patents annually on the subject. Just one of numerous examples is depicted in the drawing by George Clark of Brooklyn in his US 471267, filed in 1891. It is actually for a "Shoulder-brace" which, he says, was "adapted to draw the shoulders backwards and to throw the chest forward", but it is a good illustration of a corset—and its pained wearer.

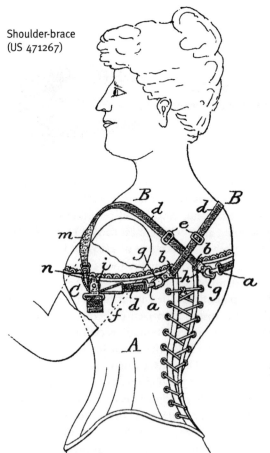

Shoulder-brace
(US 471267)

The invention of the brassiere itself has led to much argument, with Mary Phelps Jacob's first backless, elasticated effort, US 1115674 from 1914, normally being given the credit. Cup sizes were not introduced until 1935, by Warner's, with an A to D range. The B cup was initially the most popular but the Pill and improved diets have lead to the C cup being the best selling.

There were attempts to promote other undergarments, such as "union suits" or "underflannels". These were all-in-one garments. In 1874 Susan Taylor Converse of Woburn, MA, filed for her US 166190, "Underclothing for women". It was quickly nicknamed the Emancipation Suit. A gathered section across the bodice freed the bust from compression, and sets of buttons at the waist and hips helped suspend several layers of skirts. At a time of great prudity in talking (and seeing) Converse's patent talking of means "to support the breast in its natural position" was strong stuff. It was endorsed by the New England Women's Club, who sponsored talks on clothing reform by female doctors. Although the product sold well, women's groups asked her to give up the 25 cents royalty she earned on each one. Her reply was, "With all your zeal for women's rights, how can you even suggest that one woman like myself should give of her head and hand labor without fair compensation ?"

Men, too, could find union suits useful, if only to keep out the draft in poorly heated houses, and the idea of long johns continues to be popular with some. A major innovation in underwear came to the man once called the "Edison of underwear" in the middle of the night, "almost like a dream". Canadian Horace Johnson was a knitting room supervisor at the Cooper Underwear Company at Kenosha, WI. Johnson woke up his wife one night. "I want you to sew something for me", he said. At his direction she stitched multicolored cloth remnants from her scrap bag to make his vision real.

Underwear at the time had been a two-piece outfit of a button-up shirt and bulky baggy drawers which were worn all week and were bulky and uncomfortable. In winter the warmth was necessary, but it was cold when you needed to go to the toilet. There had been attempts at union suits but as one Cooper salesman put it, "Some provision had to be developed so that the seat could be opened when the occasion demanded it". The trapdoor idea was tried, but the drop seat was hard to unbutton and even harder to rebutton. The so-called open crotch type union suit, featuring strategically located button-less flaps, was no good either, as it chafed the flesh. Johnson's solution was two knit pieces that formed an overlapped X which could be drawn apart when required. He called it the Klosed Krotch union suit. In 1909 he filed for US 973200, with half the rights assigned to his employer.

The company flourished with the popular new product, and became what is now the famous Jockey International, which is still based in Kenosha. The company was also responsible for the introduction of the Jockey® Y-front® brief in 1934, which were first sold at Marshall Field in Chicago. In Britain, at least, instructions had to be given to left-handed men to wear them inside out.

Twentieth century technology has enabled many new fabrics to be created. Nylon is just one of them, and a later one used polytetrafluroethylene (PTFE) to make Gore-Tex®, the "breathable" fabric used by hikers so that their clothes no longer get clammy. In 1970 W.L. Gore and Associates of Newark, DE, applied for what became US 3953566, a process for making PTFE into a fabric. The reason for its use was that it is water-repellent. If it could be used in a porous material then the clothes would "breathe" while the water was kept out. The patent

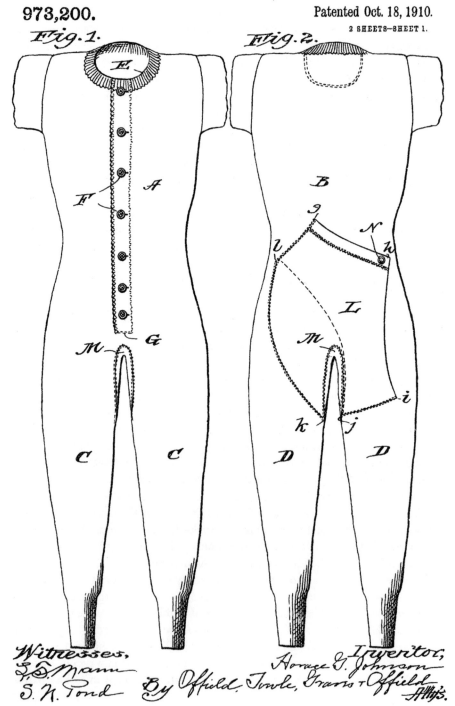

Klosed Krotch union suit (US 973200)

explains that the problem in developing it was keeping the fabric porous and of reasonable strength. Earlier methods had involved adding a "filler" to the PTFE when it was shaped into the garment and then removing the filler by for example melting or burning it out. Not surprisingly, this was costly. Briefly, the solution was to make PTFE into a paste and to shape the garment out of it. It was then stretched, heated and cooled. The porosity produced by the expansion was retained for there was little shrinking. It could then be bonded to other materials to make the final product. Few wearing the product guess that they are wearing something better known in non-stick frying pans: Teflon®, which like nylon was developed by Du Pont.

Levi Strauss's website claims that they make the only garment created in the 19th century that is still worn today. For a long time jeans were sold as hard, dark blue garments to cowboys and other Westerners. It was movies that began to make jeans a global phenomenon. *The Wild One* (1953) with Marlon Brando and *Rebel Without a Cause* (1955) with James Dean showed youths challenging "the system" and helped give jeans a non-conformist image that appealed to teenagers. They were no longer just for cowboys. The name changed too. Until 1960 "waist over-alls" was the traditional term used for jeans, but as their teenage wearers were call-ing them "jeans" Levi Strauss officially changed the name too. Nowadays, besides an interest in other colors than blue, many consumers do not want their jeans to look new. In 1993 Levi Strauss filed for US 5593072, an "Automated garment finishing system". It discussed the strong consumer demand for blue jeans that already look faded (and feel soft) when they are sold. This demand was typically met by putting them in washing machines together with pumice stones and bleach. This did not produce localised fading as seen with actual wear, and it was normal to use "workers wielding wire brushes or sanding wheels or paint sprayers with bleach or sand" to provide this. The patent was about automating the process needed to allow robots to apply this kind of finishing.

The drawing shows a conveyor belt assembly line which is operated by robots controlled by microprocessors. Loading and unloading stations (20) have means for tilting between the vertical and the horizontal position so that the clothes can move on and off. A clamp (30) locks the jeans into position on fixture (40) which includes a pair of inflatable rods to hold the legs upright and at the right dimen-sions. (76-78) is an example of a robot which might place the "worn outline of a wallet or snuff box in the rear pocket" at exactly the right place every time. Clearly, this is an important part of the American Dream to some. There have been com-plaints about excessive use of water and energy, and of pollutants, from stone-washing, and so enzymes are now used instead of stones to achieve the faded look. Enzymes can even give the blue color.

Some footwear incorporates special features. In 1993 Michael Jackson and two others applied for US 5255452, a "Method and means for creating anti-gravity illusion". This boot enabled the user to lean forward at an extreme angle without

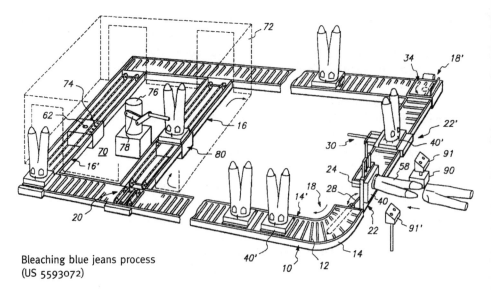

Bleaching blue jeans process
(US 5593072)

toppling over. The heels had a V-shaped slot to which a "hitch member" project-ing through a groove in the stage fitted. By simply moving forward, the boot engaged with the hitch. As explained in the patent, Jackson had formerly achieved this effect by using a cable attached to a harness round the waist. This was awkward, and could not be used at live shows (as connecting and reconnecting by stage hands was necessary). The *Moonwalker* movie, which used the leaning idea, presumably does not use this technology as it was made 5 years before the patent was applied for. If this idea sounds odd as a patent, Gregg Myles of Gary, IN, applied in 1999 for US 6101747, where a miniature basketball hoop and backboard were attached to the back of the shoe.

It is eminently part of the American Dream that youths should look cool in the latest, high-tech sneakers, even if they have no intention of exercising. This eager buying public has meant a flurry of inventions by manufacturers and others keen to cash in. Keds® sneakers were the first to be mass-marketed as canvas-top sneak-ers from 1917. For many decades when buying athletics shoes there were only two decisions to be made besides size: which color (just black or white), and whether high-top or low-top. There were a few variants, such as those made specially for basketball. This restricted choice was very boring, and there is now a much bigger variety. They also cost a lot more at generally $100 plus.

The market leader in sneakers and other athletic shoes is definitely Nike. Bill Bowerman was the track coach at the University of Oregon. He wondered why different sports did not have different and appropriate shoes, but his suggestions to the sporting goods companies were turned down. He learnt to be a shoemaker, and made custom-built shoes for his athletes. One of them was miler Phil Knight, a business major. Knight got so excited by Bowerman's ideas that in 1962 he flew to Japan to see their top sports shoes company, Tiger Onitsuka, and talked them into making his (non-existent) company their exclusive distributors in America. A company was quickly formed, they began to import, and Bowerman's designs were suggested to the Japanese and were adopted.

Times were hard. They had to pay before the shoes would be shipped, and sometimes their orders were delayed while Tiger met domestic orders. At first the shoes were sold at school and college track meets across the American West. The shoes arrived without boxes, so they were sold in plastic bags, but when a few shops were opened an opportunity was not missed. One outlet used discarded embalming fluid boxes as they "really looked nice". The name Nike, for the Greek goddess of victory, was adopted in 1968. Then in 1971 Tiger cut off supplies, annoyed at only getting a small percentage of the growing turnover. They had already set up a parallel network in the USA. Another flight to Japan resulted in a quick deal with a rival manufacturer and the flow of shoes resumed. Tiger lost the court case, and vanished from the American market.

Meanwhile technology was being developed. There was the heel wedge (against shock), the cushioned mid-sole, and nylon "uppers". There was also the major breakthrough in 1972 when Bowerman one day was using a waffle iron to make his breakfast. It suddenly occurred to him that the waffle structure could be used to make soles with excellent traction. He poured rubber latex into the iron and created the first waffle sole. Another innovation was the air-cushioned sneaker, the Nike Tailwind®, in 1979. Aerospace engineer Frank Rudy developed the concept following studies of the pressurised airbags used in the lunar landers. The trouble was that the shoes were made of poor material so that they tended to rip apart, and half the initial production was returned, but the improved Air® shoe from 1985, with visible air pockets, was a big hit. Not all of Nike's inventions are for athletes: the 1997 drawing from their US 6016613 is for a "Golf shoe outsole with pivot control traction elements".

Knight's first great marketing ploy was in 1972. He announced that "four of the top seven finishers" in the marathon at the Olympic Trials wore Nike's shoes. This was true, but the first three wore shoes made by Adidas. It was the beginning of Adidas' decline from dominance. Sales have been boosted by well-paid endorsements from superstars like Michael Jordan. The famous "swoosh" logo cost the company much less: $35 was paid to an undergraduate at Portland State University for its design. Knight continues to run the company.

The idea of sneakers lighting up at intervals when pressure is applied is sometimes said to date back to the "LA Lights" series by LA Gear, although US 1597823 by Simon Randolph of Lynch, KY, in 1925 is an early version of the concept. In 1993 LA Gear released a light Gear series with removable and interchangeable light-emitting diode (LED) cartridges. Unlike most designs the Gear batteries are replaceable. The user can turn off the LED by removing the cartridge, turning it over, and reinserting it. Earlier, designs often used mercury

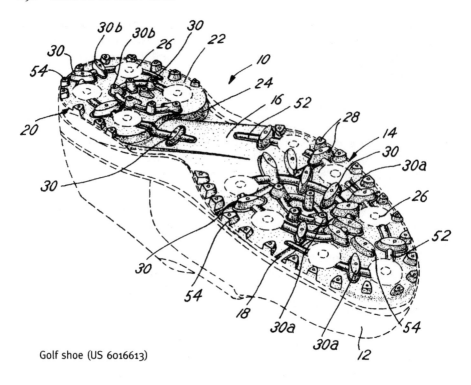

Golf shoe (US 6016613)

tilt switches, where a ball of mercury ran back and forth along a tube between electrical contacts. They were reliable, but the mercury resulted in a sneaker that was considered hazardous waste by the Environmental Protection Agency. The common replacement was a plastic tab which, depressed by the weight of the wearer, again made contact. This time they were environmentally safe, but not reliable. Plastic fatigue could occur if the tab snapped, so that the light would stay switched on until the battery died.

An example of a shoe incorporating these sort of features is US Re 37220, filed in 1993 by Carmen Rapisarda of Monrovia, CA. Flashing lights are more attractive than constant lights and make the battery last longer, but another reason for them, the patent explains, is that "More of an on and off flashing action to the light-emitting diode would provide a somewhat more flashy appearance which is beneficial from a marketing as well as a safety standpoint, by providing a more visible signal." The safety idea was that the lights would enhance visibility at night.

Figure 2 shows a wafer battery (16) which slides into the grooves (13) and (14). The two-pronged LED (17) ends in a transparent lens (18) which shows at the end of tube (22). The negative terminal of the battery is constantly in contact with the LED, but the positive terminal only connects (and turns on the light) when there is pressure by the wearer on the "weight member" (23) above it. To avoid the problem of possible fatigue—a break in the weight member—switches can also be operated by either a coiled spring reacting to pressure or a rolling metal ball reacting to gravity. Alternatively, an all-electronic method such as US 5903103 can be used to make a predetermined sequence by using an integrated circuit. A cheap and reliable solution is awaited. And all this effort is made to make the wearer look smart. Not quite so cool is the "Apparatus for simulating a 'high five'". US 5356330 is by Albert Cohen of Troy, NY, and was filed in 1993. The artificial arm pivots "when struck by a user, thereby simulating a high five". It is clearly for use by the solitary (and rather sad) fan when watching sport on TV.

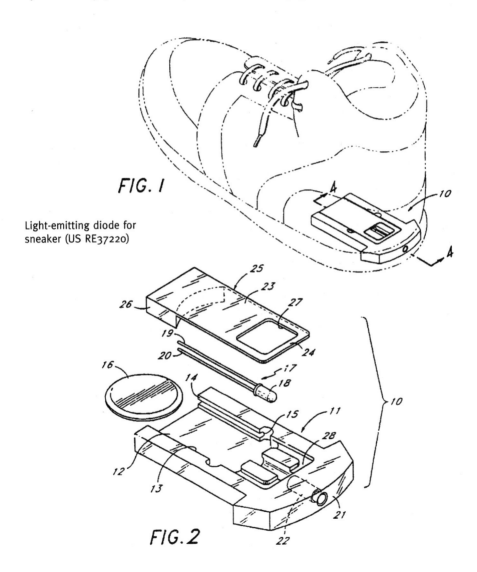

FIG. I

Light-emitting diode for sneaker (US RE37220)

FIG.2

Working towards the paperless office

··

THE Elizabethan playwright William Wycherley in *The country wife* has the line, "Go to your business, I say, pleasure, whilst I go to my pleasure, business". People can be divided into two groups: those who love their work and wish to carry on, and those who would love to retire as soon as possible. Americans work longer hours than Europeans and have less vacation. A vacation of 2 or 3 weeks is normal and there is no legal entitlement to any holidays. In Europe the average vacation is 34 paid days, with 4 weeks being the compulsory minimum. There is also a limit of 48 hours worked per week (there are, admittedly, some loopholes in this regime, as employees can opt-out). A life of long hours at work, low taxes and low benefits is condemned by many Europeans as an "Anglo-Saxon" economy—though they would quite like the higher pay and low prices to go with their own benefits. The payoff in theory is more prosperity (for some, anyway), as more effort is put in in the first place. This has translated into buying more consumer goods rather than keeping more time free for leisure.

Many inventions for the office itself have been filed. These include such innovations as a "Spiral patent office". Nicholas Bromer of Takoma Park, MD, filed in 1996 for his US 5855098. The idea was that the office building was in the shape of a spiral and you added to the end of the spiral if you needed more space, while everyone had easy access to windows. At intervals access corridors linked up the arms of the spiral. It is not clear why it had to be for a patent office: perhaps the pressure of more and more applications being sent in was thought to justify it. In 1964 Herman Miller Inc. of Zeeland, MI, introduced what they called the world's first open-plan office system, the Action Office system. It revolutionized office layout, design and the way people work. It was based on the theories of Robert Propst, a scientist who studied how workers behaved in relation to their space. He developed the theory of a "facility built on change". Propst believed that a purpose-built environment would not only motivate its occupants, but also encourage productivity. The desk would be modified to provide the right tools for the office worker rather than being standard fittings. Propst had some 60 patents for the company for different aspects of his ideas.

One of the company's patents, US 4876835, dating from 1984, is an example of the numerous permutations of modular furniture, or in this case rather partitions, in the office. The "Work space management system" was a framework of adjustable heights into which different kinds of "tiles" could be fitted. These consisted of "acoustical tiles, window tiles, work-in-process rail tiles, lighting tiles, tackable tiles, marker tiles, data display tiles, display tiles, shelf tiles, open pass-through tiles, wire management tiles, mail tiles, storage tiles, heater tiles, and cooling and air circulation tiles". Those tiles which did not fit into this apparently exhaustive list, and merely acted as separations, could be in a variety of finishes. The ability easily to adjust the layout was said to be particularly important as new factors were starting to appear, such as the increased use of computers, rapid changes in work teams and newly required standards in the office. Computer equipment had to be in fixed locations which affected the "postural, visual and social needs of the office workers". The look shown is perhaps a little soulless despite the potted plant, but then patent drawings do lack color.

A simple invention among the many innovations in office work is the pencil with an eraser at the end. Good erasers were only available from 1844 when Goodyear improved rubber compositions. Hyman Lipman of Philadelphia patented such a pencil in 1858 with his US 19783. It was not like the ones in use now: the pencil indeed had india-rubber rather than "lead" at one end, but was ready for use only after both ends were sharpened. He is said to have earned $100,000 from it, a

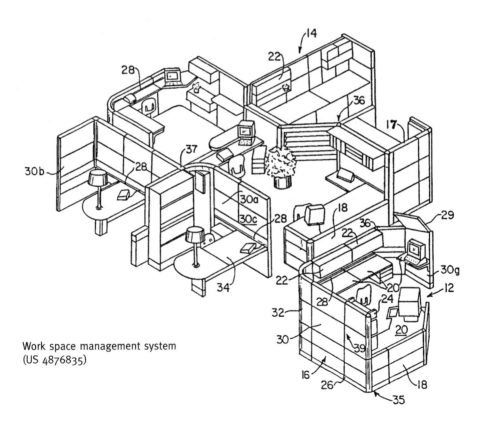

Work space management system
(US 4876835)

colossal sum for the time (and not bad today). In 1872 his patent was extended for a further 7 years, which is odd if he were making money from it, but the patent was invalidated on the grounds that you could not patent something by combining two known items. This resulted from the case of Reckendorfer vs Faber in 1875 at the Supreme Court. Joseph Reckendorfer of New York City had improved on the Lipman idea with US 36854 in 1862, which involved inserting the eraser into a slot at the end of the pencil. "The combination, to be patentable, must produce a different force or effect, or result in the combined forces or processes, from that given by their separate parts", said the Court, with convenience to the user not being taken into account.

The "window" envelope dates back to 1901. Americus Callahan of Chicago with his US 701839 patented his "Envelop" with holes which would be covered "with a section of transparent material—as, for example, very thin rice-paper—through which the sending address upon the inclosure may be readily observed, the address being so placed upon the inclosure as to register with the transparent section of the envelop". Callahan pointed out the savings in labor and that the envelope itself could be in any eye-catching color, black being a particularly good contrast to the address.

Before paperclips came along it was normal to fasten papers together with twin-pronged pins, each of which was folded back to hold it in place. Norwegian Johann Vaaler is usually credited with the invention of the paperclip, applying for US 675761 in 1901 from Christiania (now Oslo). However this was for a squared-off paperclip, rather than the rounded version normally used now. More importantly, it lacked the double oval loop used today. The truth is that the idea of holding papers together with a piece of wire had evolved over decades. But the "Gem" paperclip, which is easily the most common double-looped version used today, dates back to April 1899 at least. That was when William Middlebrook of Waterbury, CT, applied for a patent for machinery to make paperclips. He used the Gem as an example of a paperclip in the patent drawing, so presumably it was already known, although its origins remain a mystery. Within a decade this design dominated the American market. Competition continues, notably from the "Non-Skid" with its little crinkles (like bobby-pins) so that it stays in place, the "Ideal" which is used for thick wads of paper, and the "Owl", which resists getting tangled with other paperclips.

A more secure way of fastening papers together with wire is with the stapler. This useful tool did not arrive in offices until 1914. The first ones used loose or paper-wrapped staples and were hard to use: the machine had to tear out a fresh staple from the paper when in use. The breakthrough model was arguably a simplified version from the Boston Wire Stitcher Company of East Greenwich, RI, which was applied for in 1922. The idea behind US 1506073 was to prevent "crippling or buckling the staple" and ensuring that the staples were inserted "clear through the work before their legs are bent over and clinched theregainst".

A. F. CALLAHAN.

ENVELOP.

(Application filed Dec. 9, 1901.)

(No Model.)

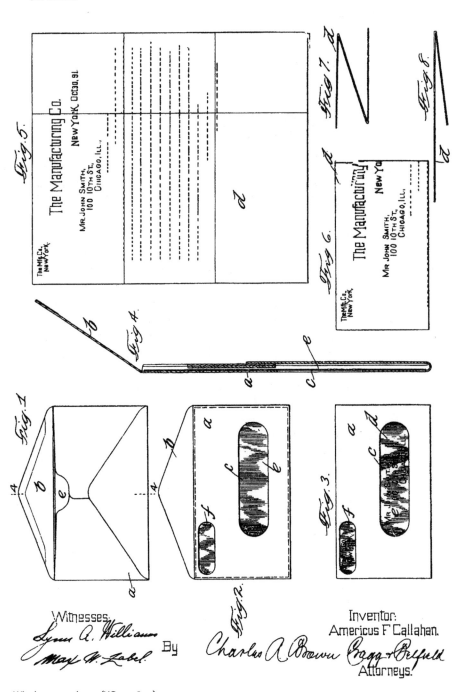

Witnesses:
Lynn A. Williams
Max W. Zabel

By

Inventor:
Americus F. Callahan.
Charles A. Brown
Cragg & Belfuld
Attorneys.

Window envelope (US 701839)

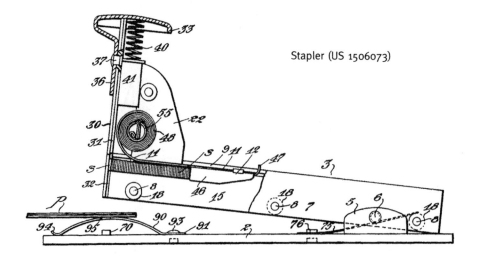

Stapler (US 1506073)

The product became so popular that the company changed its name in 1948 to that of the trademark, Bostitch®. The quick addition of the idea of gluing the staples together so that they did not have to be inserted as loose staples added greatly to their value.

All these papers have to be stored. Edwin Seibels is credited with the invention in 1898 of a vertical filing system that revolutionized filing. Businesses at the time kept papers in envelopes that were placed in rows of small pigeonholes that often lined a wall from floor to ceiling. Filing or having to find and then opening the envelopes in order to unfold the papers was very inefficient. Seibels reasoned that folding was not necessary, and that the papers could be kept in large envelopes standing on end vertically in a drawer. The Globe-Wernicke Company of Cincinnati were asked to make five wooden boxes following his idea and he applied for a patent. Seibels was told the system was only an idea and that only a device could be patented. "It was pointed out that by simply varying the size, a filing box could be made which would not infringe my patent", he later said. "Unfortunately, I overlooked the part played in setting the envelopes upright, and separating them by guide cards. This device, of course, could have been patented." Many years later, the manufacturer presented him with a bronze plaque recognizing his "pioneer work" and stating, "Business throughout the world has been helped by this idea and on it is founded an industry that provides employment for many men and women".

Telephones have been used for a long time as a convenient way of communicating. This is despite the complaint by an early potential customer from New York, when told that he would be able to speak to someone in Boston, that he did not know anyone there. By 1910 only one European in 150 had a telephone, while in America one person in 11 did. It has been suggested that the reason for the difference was that telephones in Europe were run by state monopolies, with too much to lose if the telegraph systems they ran lost business. Now people are constantly speaking to people they do not know, often to the irritation of the other party. An important advance was the introduction of answering machines. Originally answering services meant an office answering calls on behalf of others, especially doctors, who would call in to get their messages. The first working model was meant as a separate unit rather than being integrated with the telephone. It was by Kazuo Hashimoto of Japan, who first applied for the invention there in 1958, and it was patented in America as US 3376390. It could both provide an explanatory message for callers and record a message left by them. A conventional phone sat above the unit, to which it was connected by a cord. The drawing shows the contents of the unit as seen from above. Much attention in the patent was paid to circuit breakers which detected when messages ended. A few small-scale efforts, including a Swiss method in the 1930s for use by Orthodox Jews whose religion did not allow them to use the telephone on the Sabbath, had previously been invented, but this was the first to be marketed on any scale. Hashimoto was also responsible for the first digital version with US 4616110, which was filed in 1983 in Japan by his own Hashimoto Corporation. Hashimoto's work was probably the first significant invention in the office environment to come from Japan, and they have certainly made their impact felt since. Of the 35,000 American patents in classes 358 (fax machines) or 399 (photocopiers) from 1976 to 2002, about 58% came from Japanese companies. Perhaps the only non-electronic innovation in the office supplies field from Japan that has lasted is the felt-tip marker pen, invented by Yokio Horie of the Tokyo Stationery Company in 1962 (and since improved).

Offices mostly use voicemail nowadays instead of conventional answering machines. The original conception was patented in US 4371752 in 1979 by Gordon Matthews and others for ECS Telecommunications of Dallas. They enable those "If you require . . . press 1" messages which get the customer to do much of the work. Nevertheless Matthews said, "When I call a business, I like to talk to a human" (even if the human at the other end is never present during the transaction), and that is what it does by using digital (and not analogue) technology in a computer network. One of the 21 pages of drawings is shown. Figure 6 shows the hardware involved to run the system; Figure 7 shows the storage of memory; and Figure 8 is for the computer functions to run the system. Another innovation was to show the caller's details on the telephone being called. In 1983 Carolyn Doughty of Wheaton, IL, filed for AT&T Laboratories what became US 4551581 (for sending the data) and US 4582956 (for receiving the data). Hashimoto here, too, claimed to be a forerunner.

Not just words have to be sent by telephones. The idea behind facsimile or "fax" machines was known from the 19th century, and work continued on the idea of sending images over the telephone wires. However putting products such as this on the American market was blocked until 1956 because the Bell System, with a

FIG. 3.

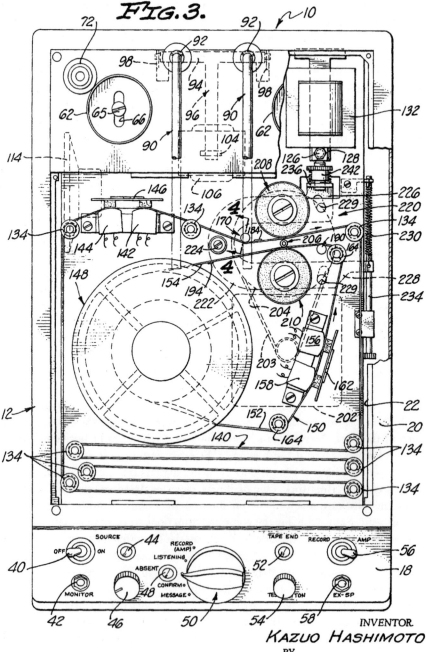

INVENTOR.

KAZUO HASHIMOTO

BY

**MAHONEY, HALBERT &
HORNBAKER
ATTORNEYS**

Answering machine (US 3376390)

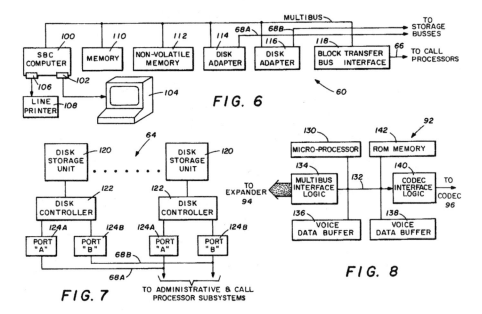

FIG. 6

FIG. 7

TO ADMINISTRATIVE & CALL
PROCESSOR SUBSYSTEMS

FIG. 8

Voicemail (US 4371752)

monopoly of telephones, prohibited the connection of non-Bell equipment to their telephone wires. They talked of "integrity": the system had to work as a whole with no interference from anyone else's equipment along the way. One day one of the AT&T lawyers passed a store window on his lunch break and saw an advertisement for a Hush-a-Phone. This was an attachment that fitted over the mouth of a telephone and let you speak quietly while still being heard. He filed a suit with the FCC claiming the Hush-a-Phone mouthpiece would cause catastrophic failure of the phone system. The FCC ruled against Hush-a-Phone, who appealed and proved that their device could not affect any telephone system. The Hush-a-Phone suit was a precedent for everyone else from then on. This and other cases led to the breaking up of AT&T into eight companies in 1984.

In 1966 Group 1 fax machines appeared. These early models cost $2,000 each and were slow and noisy. They were also only able to send or receive from identical or near-identical machines, as it was only in 1976 that an internationally agreed Group 2 standard was fixed. This is what the electronic buzzing at the beginning of receiving a fax is about: an "electronic handshake" verifies first what kind of machine is at the other end and adjusts settings if necessary, and then checks the quality of the transmission, slowing it if necessary. An example of an early fax machine is US 3475553, which was filed in 1965 by the Magnavox Company of Torrance, CA. The paper goes into the inside of the cylindrical platen (16) after the cradle (18) has been swung away from it by a lever (30). The drive motor (32)

Fig. 1

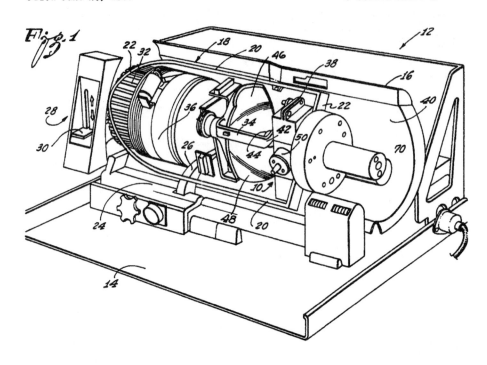

Fig. 2

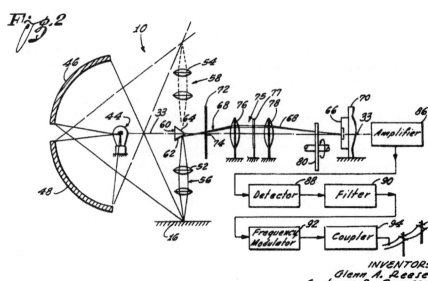

INVENTOR:
Glenn A. Reese
Gustavus B. Pearson

By Smyth, Roston & Pavitt
Attorneys

Fax machine (US 3475553)

powers the pick-up transducer (10) which uses mirrors to focus light from a lamp on the document. A yoke (34) gradually rotates to enable scanning of the whole document.

Few bought these early machines, so it was not possible to fax many others. The Japanese were particularly keen on fax machines as the old telex machines could not easily transmit their language's characters, and they played a growing part in development efforts. From 1980 the new Group 3 machines came in which were much faster–a page might take 1 minute, while the Group 1 machines would take 6 minutes–and prices fell rapidly to under $200 by the end of the decade in what had become a mass market.

The introduction of the typewriter in the 1870s meant the hiring of women on a large scale as they were thought to be faster than men. Previously one of the few office jobs to be held by women was as telegraphists. This may have been because they tended to work by themselves, and thus observed decorum, as well as the job (like typing) being reliant on speed and accuracy. Some refinements were needed such as the typewriter ribbon, which was the subject of an 1884 patent by George Anderson of Memphis, TN, with US 349026. Those of a certain age will be familiar with ribbons, and the ink getting everywhere when they were changed. More modern ribbons were divided into red and black strips to allow for color but Anderson's patent only had a red portion at the end to warn the typist that the slow movement of the ribbon off its reel was coming to an end, and that it would need to be replaced.

The typewriters themselves changed radically with the arrival of the IBM Selectric® model (US 2895584 which was filed in 1955 from Poughkeepsie, NY). Figure 5 shows an "exploded" view of the components of this "Single element printing head". Electric typewriters had already been invented by James Smathers of Kansas City, MO, in 1922 with US 1600252 (although IBM sold the first successful model from 1935). This new typewriter did not have type bars or a movable carriage. It used a revolutionary sphere-shaped element—a feature that quickly became known as the golf ball—that printed characters through a ribbon onto the paper. Typists could choose from six golf ball typefaces that could be changed in seconds. If two characters were struck simultaneously, the Selectric® stored one momentarily, then typed the other. It also used a new ribbon spooling design. To change the ribbon, a typist lifted the ribbon cartridge off the carrier and snapped a new cartridge into place. The look of the Selectric® was also popular. IBM hired industrial designer Eliot Noyes to develop a sleek and sculptured look which was patented as US D192829. The product was marketed from 1961 at $765 and quickly became profitable. A self-correcting Selectric® model was later added and by 1975 the Selectric® range had 75% of the US electric typewriter market. It was, however, soon to disappear as the personal computer began to take over, which was helped by IBM themselves patenting the first ever word processor in 1961 with US 3248705, the "Automatic editor". This patent foresaw so many features that we

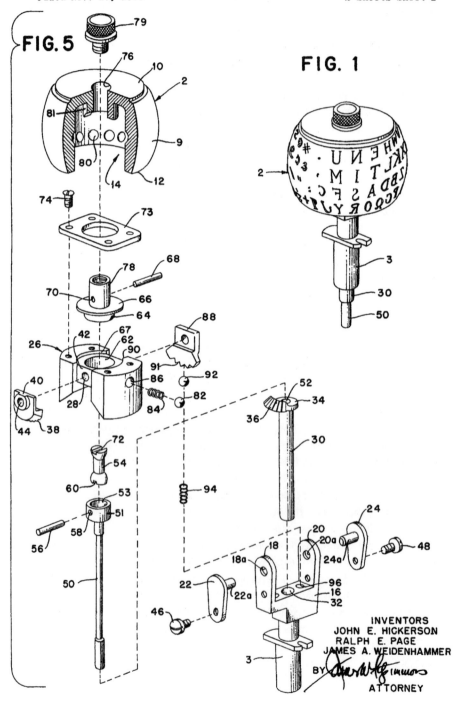

FIG. 5

FIG. 1

INVENTORS
JOHN E. HICKERSON
RALPH E. PAGE
JAMES A. WEIDENHAMMER

BY

ATTORNEY

Selectric® "golf ball" typewriter (US 2895584)

nowadays take for granted that it unintentionally makes for hilarious reading, with ideas such as erasing, inserting and moving text being carefully explained. It even allowed for networking the text as shown in Figure 1.

The Selectric® retained the old "Qwerty" keyboard (which dates back to the very first typewriter patent, US 79265 in 1868), even though there was no longer the original need to slow down typists to prevent the keys getting tangled with each other. Many attempts have been made to introduce better arranged keyboards, the best known perhaps being US 2040248 by August Dvorak and his brother-in-law William Dealey, both from Seattle, which was based on 18 years of time and motion studies. The Depression may not have been the best time for this 1932 attempt. His "Pyfgcr" keyboard had the most used letters in the middle row, with the vowels on the left. It has been shown greatly to reduce Repetitive Strain Injury (RSI) problems as well as promoting much faster typing. Dvorak died in 1975, a bitter man. "I'm tired of trying to do something worthwhile for the human race", he complained. "They simply don't want to change!" It does not take too long to learn a new keyboard, but strong pressure in favor of a particular layout is needed if change is ever to happen. A chance was missed, perhaps, when computers came into the office. As for automatic spell checkers, the first patent was US 4136395 in 1976, filed, almost inevitably, by IBM. Many an office worker would be lost without a more modern version of the "System for automatically proofreading a document".

Another electronic innovation was the pocket calculator. Texas Instruments of Dallas began work on a hand-held calculator, codenamed 'Cal-Tech', to show the potential of their recently developed integrated circuits, which were not selling as well as they had hoped. These were much smaller than transistors, but were more expensive and the market thought that they were just new and unproven. Jack Kilby, who had invented silicon chips for the company in 1959, was one of the three inventors of US 3819921, a "Miniature electronic calculator", applied for in 1967. It had just the four basic functions (add, subtract, divide, and multiply), was battery-driven, and even had a thermal printer, presumably as printouts were seen as the conventional way of getting results from a computer. Up to 12 digits could appear on the display (1) (which incorporated a magnifying lens), with each digit showing as soon as that part of the calculation was performed. Such calculators were known, but not in this small size. Its dimensions were proudly stated in the patent as being less than $5 \times 7 \times 2$ inches, and it weighed 45 ounces. By pressing the keys contact was made with circuits (6) to encode the command for the logic circuits making the calculations at the bottom of the device, next to the battery. In between was the spool of thermal paper, with its tiny printer at (4). It is difficult now to appreciate the revolution which such a machine symbolized, although this was only a prototype which never went into production. Samples were shown to a number of companies, including the Japanese company Canon, which saw the potential. Their Pocketronic model was the first pocket calculator to go on the

FIG. 1

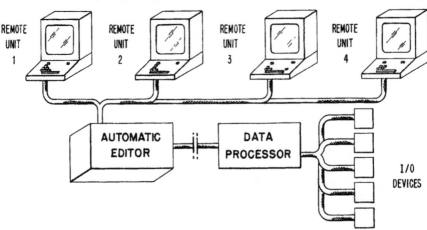

FIG. 2

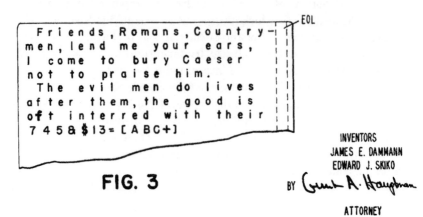

FIG. 3

INVENTORS
JAMES E. DAMMANN
EDWARD J. SKIKO

BY _Grant A. Hauptman_

ATTORNEY

Word processor (US 3248705)

market in 1970. It was priced at a hefty $400 and had many similarities to the "Cal-Tech", including the thermal printer (which was made for them by Texas Instruments itself), although that provided the only display of the results. Prices soon fell as demand increased, and nowadays most pocket calculators are slim, solar-powered devices, many almost as small as a credit card, although they do lack printers. They are also mostly made in the Far East.

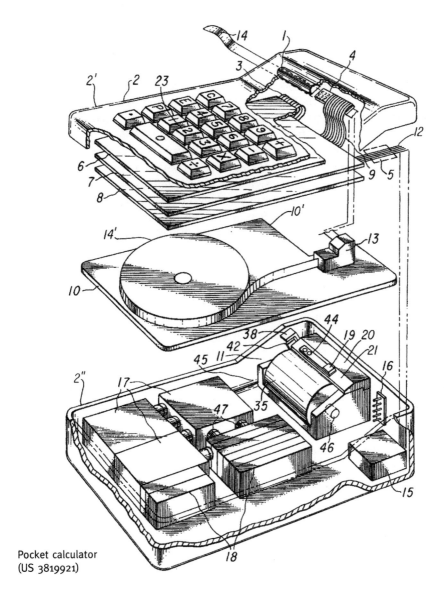

Pocket calculator
(US 3819921)

By this time the computer age was on the horizon. Many think of computers from the point of view of their own machine at home, but business use and needs, are in a way far more significant in driving change. IBM was heavily involved with the creation of the idea of personal computers rather than huge machines sitting in specially air-conditioned rooms. It was Thomas Watson, the Chairman of IBM, who is alleged to have said in 1943, "I think there is a world market for maybe five computers". Fortunately for his company he was wrong. Computers at first were machines which provided tape, printouts and so on recording the answers to mathematical questions. There were no screens where you could draw or type. As each computer had at least one dedicated attendant they were certainly not "personal".

Strictly speaking, the "PC" is the Personal Computer as manufactured by or for IBM, but most nowadays are clones made by other manufacturers. The first IBM PC, introduced from 1981, was meant solely for word processing or for financial spreadsheets. It was not the first to have a 16-bit microprocessor (which made it more powerful for processing data), but it did make it easy to convert programs from the old CP/M operating system to run on it. The IBM system was open, meaning that they only owned the rights to the keyboards and the ROM-BIOS chip. This helped the growth of the concept, as within a year other companies began to market clones, with redesigned keyboards and reverse-engineered chips. Reverse engineering is when the principle behind something, usually software, is explained to engineers who then design or write something that does the same, hence avoiding violation of copyright. The more PCs there were the more software was written for it, and hence more machines were bought, in a virtuous cycle.

IBM would have benefited hugely by having rights to the operating system for their PCs and then licensing it out to all who asked. When they wanted a system designed they approached Bill Gates who had a small Seattle company called Microsoft. They in turn purchased the rights to QDOS (the Quick and Dirty Operating System) from its authors for $50,000 without mentioning that IBM were interested in the enhanced version that they intended to write: MS-DOS. IBM was talked into allowing Microsoft to keep the copyright to MS-DOS. Microsoft based its huge empire on this highly lucrative deal, with some 90% of the world's PCs now running a version of Windows®. It has also achieved a near monopoly in the more common kinds of applications such as emailing, as everyone is very well aware, and many are angry about. While this restricts choice (including the practice of putting lots of applications into one bundle, and preloading the PCs), the wide prevalence of the software does make for ease of use, as the system is mostly similar at different PCs or sites as the user moves around. Microsoft also tries to give a family feel to different applications so that you can often guess how to use a new application. Personal computers have been greatly helped by the huge growth in size and cheapness of computer memory, both short-term RAM (great for fast-moving games) and long-term (ROM). Yet few know how to get the most, or even a lot, out of their computers. Many companies will spend millions on computer systems but much less on training their staff in how to get the most out of them.

Traditionally, the software to operate the computers relied on copyright protection—a right against copying. It protected the code in the programs rather than any concept represented by it. Thus one way of say creating a drop-down window could be protected. In practice reverse engineering flourished, where the program would be analysed and its functions described to a team who would then write code to carry out the same task, avoiding copyright violations. The long protection given in copyright (now lifetime plus 70 years) was irrelevant as few programs last for more than a few years. Patents last only 20 years, but their protection is much stronger. The first patent for software is thought to be US 4270182 by Satja Asija of St Paul, MN, for text retrieval.

The court case which opened the floodgates for software patents was the "State Street" case. US 5193056 was filed in 1991 by Signature Financial Group of Boston, and was called "Data processing system for hub and spoke financial services configuration". It was a system for allocating gains and losses among a family of mutual funds that were pooling their assets and gives a flowchart but not the code used to carry it out. A 1998 Federal Circuit case ruled that such inventions were patentable. The objection by State Street Bank had been that the idea could only be covered by copyright. The court ruled that if an algorithm produced a useful, tangible and concrete result, it could be patented. By allowing such patents the entire concept (and not the specific code) is barred to others.

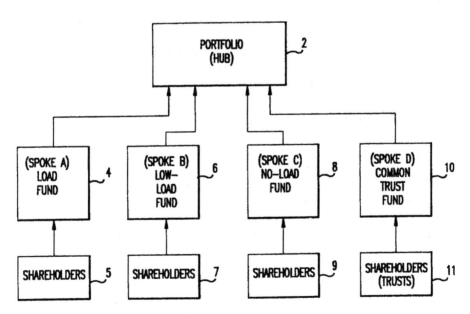

Business method patent (US 5193056)

Innovation in software design could become a nightmare of avoiding patents, or obtaining licences after identifying prior work in the area.

These "business method patents" have become very popular, and over 7,400 had been allocated to subclasses within Class 705, "Data processing: financial, business practice, management, or cost/price determination", by the end of 2002. The oldest, US 3931508, dates from 1973. It is a German-origin patent for a taximeter.

Jeff Bezos, the CEO of Amazon who has himself been criticized for patenting their one-click ordering system, has suggested patent reform, with only 3 to 5 years' protection for software patents. This would reflect the fast development cycles in the industry and would permit concepts to become available to the public in a relatively short period of time, while still keeping an advantage for those first in the field.

Above all, perhaps, in software looms the promise if not always the delivery of the web. The internet evolved from a closed network of connections for use in any future war. It would allow alternative telecommunications routes or computers to be used if necessary. The Advanced Research Projects Agency had been formed in 1957 in response to the perceived threat of Sputnik and set up ARPAnet, a primitive internet, in 1969, with email coming along 2 years later. Many ideas in both hardware and software were worked out or adapted by many different persons or organizations to contribute to what became the web. The web as we know it came out of work by Tim Berners-Lee, a British software engineer, adapting the idea of the ARPAnet. He has said in his *Weaving the web* (1999) that "Inventing the World Wide Web involved my growing realisation that there was a power in arranging ideas in an unconstrained, weblike way". By 1980 his first thoughts took shape of a computer program that would create a space in which anything could be linked to anything, with computers communicating with each other. This single global space would have bits of information labelled with an address so that related bits could be retrieved. Human effort would be needed to pose the questions, but computers would do the hard work of gathering the information together.

Berners-Lee did not realize it, but others had already been thinking along similar lines. In 1965 Ted Nelson, who was taking time off from his sociology masters at Harvard, wrote of something which he called *hypertext* within a project called Xanadu, in which links could be followed from say a quotation to get more information. Tiny payments would be automatically made by the user to the creator of the information. Doug Englebart, inventor of the computer mouse among other things, was also working in the 1960s with primitive email and hypertext links by using a five-key keyboard. A major problem was the vast cost of computing power but the arrival of the personal computer, and cheap memory and processing, was what enabled the internet to be created. While working on a contract at CERN, a physics research facility in Switzerland, Berners-Lee began working on a program called Enquire, where bits of information were linked to each other. Some were internal links within a document, and others were external. Berners-Lee wrote the first HyperText Markup Language (HTML), with commands built into plain (i.e. ASCII) text so that the software could both present the information and allow searching for it. This common language is essential because of different software packages (and versions of them), although problems do occur in some cases, notably when new versions keep on coming out and users have to scramble to buy the software. Ironically considering that an engineer wrote it, HTML is not good for scientific and mathematical formulae, and work continues on versions which will improve their presentation. Work is also being done on making concepts more easily findable such as using new mark-up systems. The ability to find images on the Google search engine is a simple example of their value.

CERN in 1993 declined any intellectual property rights in the web, which saved expense and licensing for everyone. Soon Americans had taken Berners-Lee's ideas and run with it, with a world of software (as well as concealed hardware) springing up in support. The web has become a virtual library of knowledge and thoughts and hence a part of the American Dream, as it is such a wonderful way to present or to look for information—text, maps, images or whatever. *If* you know what you are doing and where to look, the web is often much cheaper and easier to use than it is to go to libraries or to contact an agency. Those too poor to have access, or who do not understand how to use it, suffer. To be information-poor is often to be poor in wealth as well in today's Information Society.

People are worried about security on the internet. It is a great place in which to sell (no premises are needed, few staff, and you are open all year round), and convenient in which to buy (no need to go out, no annoying sales staff), but it is essential that people can safely give credit card and other details. Those working in the field, when discussing methods of concealing data, talk about Alice and Bob exchanging information without Eve overhearing them. Alice and Bob are not acquainted (the usual scenario in selling situations), so how do they use a cipher that only they will know? The standard answer had been to pass the secret data (or "key") that enables encryption over by hand, but that caused enormous logistic problems, and was useless if people did not know each other. This fascinating problem, which puzzled experts for years, would take far too long to explain here in detail.

The solution used routinely on the internet is given in US 4405829, which was applied for in 1977 by three MIT staff members after years of research. It is called the RSA cryptographic method after the initials of their surnames. There are two keys, one which is known to all, the public key, and one which is kept secret, the private key. The public key is the product of two huge prime numbers multiplied together, and the private key is those original numbers. It would take a vast amount of computer effort for others to identify the original numbers and hence decrypt the message. In 1997 it was announced that this problem had in fact been solved in 1973 by Clifford Cocks, a mathematician who was then a new recruit at GCHQ. This is the British establishment which covertly listens to encrypted broadcasts in

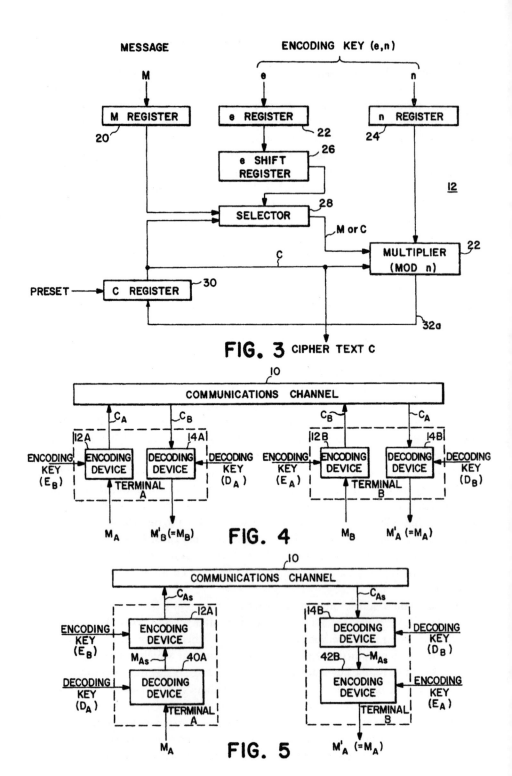

MESSAGE

ENCODING KEY (e,n)

M

e

n

M REGISTER — 20

e REGISTER — 22

n REGISTER — 24

e SHIFT REGISTER — 26

SELECTOR — 28

12

M or C

MULTIPLIER (MOD n) — 22

C

PRESET → C REGISTER — 30

32a

FIG. 3 CIPHER TEXT C

10

COMMUNICATIONS CHANNEL

C_A C_B C_B C_A

12A   14A   12B   14B

ENCODING KEY (E_B) → ENCODING DEVICE   DECODING DEVICE ← DECODING KEY (D_A)   ENCODING KEY (E_A) → ENCODING DEVICE   DECODING DEVICE ← DECODING KEY (D_B)

TERMINAL A   TERMINAL B

M_A $M'_B (=M_B)$ M_B $M'_A (=M_A)$

FIG. 4

10

COMMUNICATIONS CHANNEL

C_{As} C_{As}

12A   14B

ENCODING KEY (E_B) → ENCODING DEVICE   DECODING DEVICE ← DECODING KEY (D_B)

M_{As} 40A   42B M_{As}

DECODING KEY (D_A) → DECODING DEVICE   ENCODING DEVICE ← ENCODING KEY (E_A)

TERMINAL A   TERMINAL B

M_A $M'_A (=M_A)$

FIG. 5

RSA cryptography (US 4405829)

foreign countries. He took half an hour to work out the solution after it was casually mentioned to him as what he assumed was a trivial problem. GCHQ did not
patent it for two reasons: they thought, erroneously, that it was unpatentable, and
they wanted to keep it secret anyway. A problem in the area of encrypting messages
is that those using emails often need (or at least want) to keep their communications, and not just financial transactions, confidential. Secure methods are a boon,
though, to criminals and terrorists as well as to businessmen. It was found for
example that the Aum Shinrikyo sect, which carried out gas attacks on the Tokyo
subway in 1995 (leaving 12 dead), used RSA to encrypt some documents.
Governments would like "key escrow" where the keys are kept by them for use in
case of national emergency. There is no easy answer to the dilemma between the
rights of privacy and the need for security.

There are also problems with royalty demands. A claim for royalties came out of
a way of speedily accessing images. People do not want to wait minutes for images
to appear on the screen. JPEGs are frequently used to store images compactly for
rapid access and do so by removing redundant pixels. The Joint Photographic
Experts Group worked out the technology in 1991 and made it available for all.
Meanwhile US 4698672, entitled "Coding system for reducing redundancy", had
been filed in 1986 by Compression Labs of San Jose, CA. Forgent Networks of
Austin, TX, acquired the rights to this patent in 1997. A newly appointed CEO,
Richard Snyder, had a search done through the company's patents which revealed
the invention. In July 2002 they claimed rights in the technology. The company
had been a maker of videoconferencing hardware with declining revenue, and had
decided to change to being a video technology firm focusing on software and
patents. This move was stimulated by a patent deal 2 years earlier that had resulted
in $45 million of revenues. Forgent has persuaded two companies to sign its licens-
ing agreement. The patent is likely to be contested—unless the cost of litigation
prohibits it.

Hypertext links are now taken for granted but here too there was a request for
royalties. British Telecom, on reviewing their patents, came across their US
4873662. Although this patent was filed in 1976 it was not granted until 1989,
which meant that it would not expire until 2006. In June 2000, after noticing it in
a review of their own patents, they announced that they intended to ask for royalties. A New York judge in rejecting the claim in August 2002 said that the language
of the patent was like old English, and that comparing a 1976 computer to a 2002
computer was like comparing a woolly mammoth with a jet aircraft. People who
use the web often expect it to be free, and to be able to use it for as long as they like
for the fixed cost of a local call (not true in most of the world, where you normally
pay for the time you take). Perhaps this is particularly so among the new generation who have grown up with it. The trouble is that *someone* has to pay for the cost
of it all, especially if claims for royalties on software and hardware have to be met.
There is only so much revenue that can be earned by charging companies for

pop-up ads, directing people to a site and the like. Increasingly, websites that are useful for business are becoming priced subscription sites.

Offices have not become paperless despite all the hype about computers. Trees are still being cut down at a tremendous rate. This is partly due to the continued survival of "analog" publishing (printed books and so on) as opposed to digital publishing, but also to a constant desire by many to read and keep printouts. This is less feasible when on the move. A possible solution to keeping in touch with the office has been offered in Microsoft's Smart technology, announced in January 2003 at the Consumer Electronics Show, Las Vegas. Smart Personal Objects Technology™ (SPOT) will be a "baby" version of Windows® which will add intelligence to things such as wristwatches and pens. Smart Displays are simpler still, being touch-sensitive displays, again with wireless links to a PC. The unveiling of the new technology featured a snazzy wrist watch waved enthusiastically by Bill Gates which can tell the latest news and weather besides the time. Fridge magnet versions will follow. Then the office will truly be with us all 24/7. To cope, a stress reliever such as that offered in US 5195917, the "Tear-apart stress relief doll and method", might be appropriate. This invention by Mary Russell and Cheryle Karnikis, both of Washington state, dates back to 1989. Velcro® is used to keep the different parts together before rage occurs. The patent has a sports referee as an example but your boss might be more appropriate.

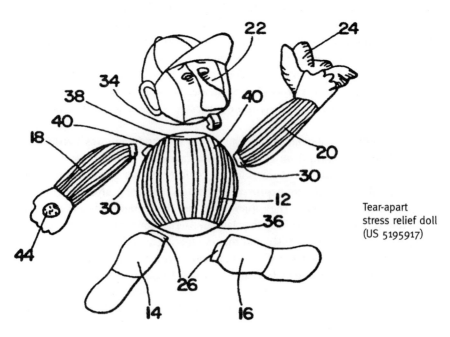

Tear-apart
stress relief doll
(US 5195917)

Truth, justice and the American way

....

So far we have seen many examples of life being enhanced by patented inventions. Yet none of it amounts to much if your possessions cannot be defended against criminals or other enemies.

The simplest way is to use a lock. Locks date back at least 10,000 years, to Iraq, but more advanced locks were made by Joseph Bramah in the 1780s and Jeremiah Chubb in the 1810s, both in England, which introduced sliders and bolts. These needed "fat" keys, and the first successful thin, flat key working a pin-tumbler cylinder lock was US 31278 by Linus Yale Jr of Philadelphia in 1861. There were also numerous early burglar alarms, including US 8439 from 1851 by John Bolen of New York City. This was the first of 167 allocated to class 116/87, "Detonating alarms". Modestly described by the inventor as possessing little apparent novelty, it was for tripwires which led from windows or doors to several guns. It was not, he said, "intended as a destructive weapon", but rather as an alarm. In the same vein but with a more deadly purpose was US 60960 from 1867 by Pete Swisher of Versailles, OH. It was a "compound gun" which was rigged to fire simultaneously in different directions. "The explosion", we are told, "will alarm the household and probably shoot the burglar". On the other hand, the "Burglar-trap" from 1868, US 77582, provided a trapdoor which like these two had to be set by the householder before going to bed. It was clearly not just the burglar who was at risk from such contraptions. In modern times, the idea of a complete home security system incorporating television surveillance is normally attributed to African Americans Albert and Marie Brown of Jamaica, NY, with their US 3482037, filed in 1966. It is so clear that it requires no comment.

To prevent car theft we have car alarms, but sometimes the car is deliberately immobilized by the police. The famous "Denver Boot" was invented by Frank Marugg, who besides being an inventor was a musician with the Denver Symphony Orchestra. The Sheriff's Department came to him to ask for help with their parking enforcement problem. They had decided that they needed a device to immobilize vehicles whose owners did not pay their parking tickets. In 1955 he filed for what became US 2844954. Various websites claim that three other Denver residents had the idea. It is often said that the original idea was to prevent theft of the car, but Marugg's patent clearly states that it is "a clamp that can be used by officers of the law to impound cars", with its familiar horizontal bar extending from the hubcap to the side. Adaptations are now needed as in the 1950s most cars had similar wheels.

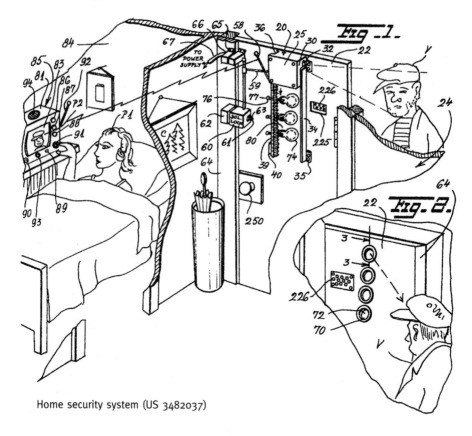

Home security system (US 3482037)

To prevent theft itself, the LoJack® car-tracking system is widely used. It is the only system with direct connections with law enforcement officers. The LoJack Corporation of Boston applied for US 4818998 in 1986. A transponder in the car acts as a radio receiver until activated by a signal when it silently transmits signals, which can be homed in on by a network of ground stations to help the police recover the vehicle. The system uses the frequency 173.075 megaherz, which has been permanently assigned to it as a police frequency since 1992. The drawing's mention of the FBI's NCIC computer refers to verifying if a stolen car is equipped by querying the database and then automatically causing the stations in the same district to send activating signals. The transponder is powered by the car battery but in case that is disconnected there is a small, rechargeable battery. A recovery rate of over 90% is claimed. The National Bureau of Economic Research estimates that equipping just 2% of the cars in an area with the system can reduce auto theft by up to a third by helping the police catch the professional criminals who commit many of the thefts. There must be apparatus in the area to pick up the signals, and to avoid this many police aviation units track cars from the air.

Some criminals do not bother to rob banks, preferring to pass off counterfeit currency, checks or credit cards. The passing off of counterfeit currency is a real problem, aggravated by the fact that American bills are all the same size and rarely change their design, unlike normal European practice. The Treasury is confident that counterfeiting is not possible. Nevertheless there are numerous patents for methods of monitoring for correct currency. New models of color photocopiers are routinely checked out by Treasury officials across the world to see how well they can duplicate their currency. Often they provide very good copies. What is claimed to be the only counterfeit detector pen, Smart Money™, was designed to help store clerks to detect fakes. It contains an iodine solution that reacts with the starch in wood-based paper to create a black stain. When the solution is applied to the fiber-based paper used in real bills, no discoloration occurs. The pen can only detect bills printed on normal copier paper instead of the papers used by the US Treasury. This invention dates from 1990 and was patented as US 5063163 by Ach Group of Tampa, FL.

Some prefer to get their money by robbing banks, and this has inspired many suggestions for capturing robbers such as US 3313250, for a trapdoor, US 3965827 where a huge container drops down over the criminal, and US 6219959 which drops a net. Many focus on placing devices within currency packs to set off tear gas or to release a dye. A 1930 example of this is US 2041577 by Clarence Sutherland of Dayton, OH. The patent commented that there were a number of

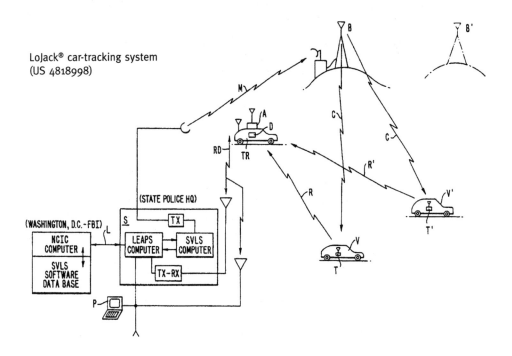

LoJack® car-tracking system
(US 4818998)

devices for releasing tear gas from walls, ceilings, watchman's night sticks or fountain pens, but all had to be activated by someone. This one was automatic. Two gas
shells (such as, he suggested, shot gun shells) would be held in place by clips, and
a mercury tilt mechanism activated a switch powered by a dry cell battery that
would activate electric caps on the shells if the pack were lifted by the robber.

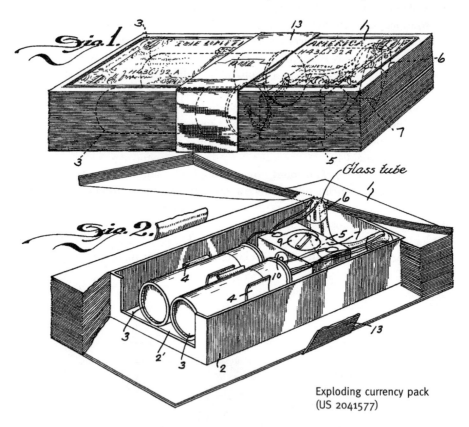

Exploding currency pack
(US 2041577)

Presumably the teller would refrain from handling the bundle. The idea of dye
being released from the pack dates back to 1960 with US 3053416, when a teller
had to push a button to activate it. To avoid angry confrontations as the pack
exploded in the robber's hands, it is normal now for electromagnetic fields at the
exit to trigger the mechanism in the bundle after a time delay. This was introduced
by US 3564525 in 1967 by Harold Robeson of Atlanta. However, that technique in
turn had its own problems. A company with an obvious interest in disabling robbers, Mace Security International of Bennington, VT, explained how it worked in
US 5952920 dating from 1999. A small metal container normally contains the
chemical inside the pack. But the criminal fraternity quickly got wise to the strat-
egy. "Thieves know this, and will often either fan the currency packs to see if they

behave as regular currency would when being fanned, or if they are stiff and unbending. Thieves have been known to bang the currency packs against the teller counter to see if they are soft, or if they contain a hard metal canister."

Their solution was to sew actual currency bills together loosely so that the edges could be fanned; splitting the chemical pack into smaller segments at opposite ends of the currency pack so that it would bend and not feel heavier or stiffer than a real currency pack; and adding foam cushions so that the pack did not make the wrong sound when banged on a hard surface. The pack would release dye to stain both the currency and the thief, emit a cloud of smoke, and (where allowed by state law) emit tear gas. The bank doors would also automatically lock once the thief left the premises. A coded transmitter/receiver mechanism automatically activated the bank's silent or audible alarm when the pack was removed from the drawer.

Sometimes the intention is not theft but rather assault. One kind of assault is putting foreign bodies of one sort or another into candy at Halloween. US 5735548, "Food donator identification container", offers a solution. It was filed by Gwendolyn Anderson of Newark, DE, in 1994. She talked of the use of X-ray machines in hospitals to screen candy for "foreign objects", but this was expensive, did not help with tainted candy, laid hospitals open to liability suits, and did not identify the guilty party. The idea was that the recipient wrote down the name and address of every donor on special bags as it was handed over. The illustration shows the bags lying within an outer bag. This invention had excellent intentions, but it is unlikely that many children collecting from the door, let alone the grownups handing out the candy, would comply with its requirements.

Protection from physical assault is understandably a matter of concern to citizens. Sometimes overtures to the assault are subtle. US application 2001/0046710 by Charlyne Cutler of Las Vegas is for her Guardian Angel Test Kits. This patent, almost uniquely, explains how the invention was thought of. "The idea of the Narcotic detecting Test Strip for a beverage arose when Charlyne Cutler, sole-inventor, was watching a crime report on GHB Date Rape. With convenience and economics in mind, using a small piece of paper (test strip) or such, to detect the drug in a beverage came to mind. The idea was written down and contemplated. Without a chemical background Charlyne consulted with many scientists to create C-3, a solution which consistently responds in the presence of GHB". A problem with this might be explaining to your date why you are putting a piece of paper in your drink and then examining it.

Among defensive suggestions, there are hundreds of patents for body armor, some intended for battlefield use. Presumably for civilian use were US 6453791, "Concealable body armor briefs", and the three patents filed during 1991–93 by Oliver North of Contra Hearings fame and others for a "Lightweight ballistic protection device" (US 5327811), "Ballistic shield" (US 5377577) and a "Removable ballistic resistant armor seat cover and floor mat" (US 5448938). These patents were assigned to Guardian Technologies International of Sterling, VA. Sometimes

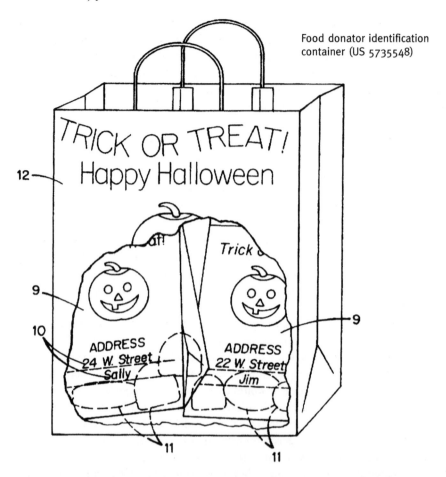

Food donator identification container (US 5735548)

only one article is armored, such as a clipboard (with a concealed weapon underneath it) in US 6453791, which is by no means the only one of its kind.

Some inventions are intended for direct action. US 4698844 offered the "Anticrime spike needle". A spike was concealed in the clothing with a steel wire wrapped around it. "When a pressure greater than normal or friendly" was made, the wire came away and the point was exposed to "discourage and even stop any further overt act". Some defenses have to be more clearly carried around by the owner. US 5570817 is for a palm held pepper spray, while US 5859588 is more versatile, being a purse which contains both a pepper spray and an "air siren". A button on the handle activates them so that the spray is dispersed through a vent. The carrying of guns is an American's birthright, and in some states a concealed weapon can be carried. In 1912 Leonard Woods of St Louis, in his US 1073312 described a pocket watch that was actually a miniature pistol. The patent stated, "It may be presented and fired at a highwayman while apparently merely obeying his command to 'hand over your watch and be quick about it!'". The watch stem was actually the barrel. There was also US 1474292, the "Walking stick machine gun"; US 2844902, a "Fountain pen pistol"; US 3554570, a pistol hidden in a flashlight; and US 5782025, a "Concealed buckle gun". There were also several that used pencils, such as US 5062230. Not all guns are concealed, with hundreds of patents for holsters, as well as oddities such as US 2423448, a "Fist gun", where the gun was mounted on a glove.

Kidnapping has resulted in many patents offering solutions to tracking people down. B.I. Incorporated of Boulder, CO, in 1986 suggested in their US 4885571 a tag attached to the ankle. It was intended for those under house arrest, and included anti-tamper technology, but the same concept could be used to track abducted persons (or mental patients or pets). US 5079541 is for an "electronically and visually detectable diaper" to prevent the theft of babies from hospital, while US 5629678 offers the rather drastic solution of an implantable transceiver which could be remotely actuated, or actuated by the implantee. The movement of the muscles provides power to the battery, and a biological monitoring feature allows the device to summon help in a medical emergency. The problem with many solutions is that they do not warn the criminal that they will be tracked down. It is surely better to deter kidnappers in the first place by publicizing that there is protection.

The idea of a leash connecting a child to a grownup has not had "substantial commercial success", commented US 6263710 by Protective Solutions of Raleigh, NC, dating from 1998, with its own interesting way both to publicize the protection and to prevent an abduction in the first place. The child wears bracelets on each arm. If an abduction is threatened, the child wraps arms around a tree or something similar and the bracelets lock like handcuffs. Care would have be taken when washing hands, or hugging a parent. An earlier method was US 5815467 for a "security device" attached by a wristband to the child. It held two chambers, separated until the device was "armed". This would mix carbon dioxide or another suitable gas with a harmless spray. The child would presumably know when to use the "activation member". The device was preferably disguised as a watch.

There are many patents for preventing airplane hijacks. In 1970 there was US 3680499, which was for a special telephone outside the cockpit. "When a hijacker identifies himself" the stewardess requests that he speaks to the pilot on the telephone. If the pilot so decides, a button is pushed which releases from the telephone a knockout gas. Just as obliging, it seems, would be the hijacker in US 3841328, from 1972. It was for a "passenger disabling apparatus" mounted under an airplane seat which was remotely actuated by a pilot or crew member. A hypodermic injection apparatus drove a hypodermic syringe needle through the seat cushion to "sedate or kill the passenger". Most of the other patented methods involve methods of keeping passengers out of the cockpit. The problem with this is that the crew are likely to be kept out as well, and there may be unfortunate consequences if the pilots fall asleep as a result. This is apart from threats to those outside the cabin.

After the crime is committed, the police have to capture the criminal. For many years the police have employed artists to sketch witnesses' descriptions of suspects. A famous attempt to improve this was the Identi-KIT® system, the patent for which was filed in 1957 by Hugh McDonald. He had spent many years working in the Identification Bureau of the Los Angeles Police Department, and dissected 50,000 photographs to work out the main facial attributes. He ended up with 37 noses, 52 chins, 102 pairs of eyes, 40 lips and 130 hairstyles—as well as headgear, moustaches, and so on. It was first used to apprehend a criminal in February 1959. A liquor store in Los Angeles had been held up, and the owner was able to give a clear description to police. From this a picture was assembled and circulated in the neighborhood. A suspect was named, who confessed as soon as police approached him. Transparencies or "foils" of the different features were coded by mnemonic category such as A for age lines and S for scars and were stored in separate compartments in the portable kit. A catalog was used to help the technician decide which foils were to be used as overlays to build up the faces. The foil numbering, besides helping keep them in order, meant that a list of the codes used could be sent to other locations. Examples of the codes are given below the suspect's picture. The grid was to help place scars (as shown on the suspect). Although ingenious, the system was cumbersome. Sketches by a police artist are often preferred over the photographic or computerised versions as well, as many feel that sketches are more useful as people tend to see the face as a whole. Recalling is difficult, while recognising is easier. To overcome this problem the University of Stirling in Scotland has designed the computerised EvoFIT, which present 18 random photographs of faces to the witness, who picks the few which are most similar to the suspect. The system then looks for another 18 random selections which share those characteristics, and so on, until a face is agreed on.

With the suspect apprehended, he or she may need to be restrained. There are hundreds of patents for batons, clubs, nightsticks and the like. These include US 1986682 from 1934, a "humane police club" which could "stun a criminal without danger of fracturing the skull or cutting the scalp". It was made of rubber rather than of the usual wood. Of course the classic way of restraining the prisoner is with handcuffs. The Bean handcuff company was a leading company in the 19th and early 20th century. The Bean handcuffs featured a unique release button that locked the cuff. This allowed a police officer to carry the cuffs closed, but unlocked, a feature that no previous handcuff had. This dates back to 1884 and US 308075 by Iver Johnson of Worcester, MA, who was working for the company.

The run of the mill criminal is not the only threat to public safety. Sometimes a nation must go to war to defend itself, and this is inextricably mixed with patriotism in, particularly, design patents. There are many inventions for armaments of all sorts (though some are presumably kept secret), but some designers also show

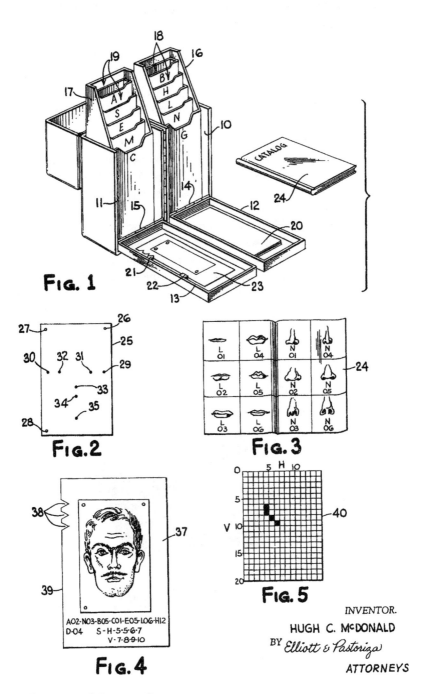

FIG. 1

FIG. 2

FIG. 3

FIG. 4

FIG. 5

INVENTOR.

HUGH C. McDONALD

BY *Elliott & Pastoriza*

ATTORNEYS

Identi-KIT® apparatus (US 2974426)

their love of their country and disdain for the enemy by using design patents for the appearance of pins, flags and other goods. Design patents, which cover appearance rather than function, began in 1842, and there is a published list of titles of designs for 1842 to 1873. This includes many medals, medallions and statues, some of which give a name in the title. Oddly, from these named titles there is just one for George Washington, while there are 16 for Lincoln, all but one dating from 1865. The exception is from 1860 when Leonard Volk of Chicago applied for US D1250 and (hedging his bets) US D1203 for Democratic rival Stephen Douglas. The Lincoln design was apparently based on the bust he made before the Presidential Election.

After Lincoln in the listings come Ulysses S. Grant and Daniel Webster with four each, with politicians and generals being well represented. The only one for an African American was US D3151 for Frederick Douglass, which dates from 1868. The sole design listed for George Washington is US D5150 by James Miller of Philadelphia in 1871, shown here.

Wars have, sadly, been the reason for many patriotic design patent registrations. The Spanish-American War started after an explosion on the USS *Maine* sent it to the bottom of Havana harbor on February 15, 1898 with the loss of over 200 lives. It was assumed that the Spanish, who in the face of American disapproval were brutally suppressing a rebellion in their Cuban colony, had deliberately

The Father of his country (US D5150)

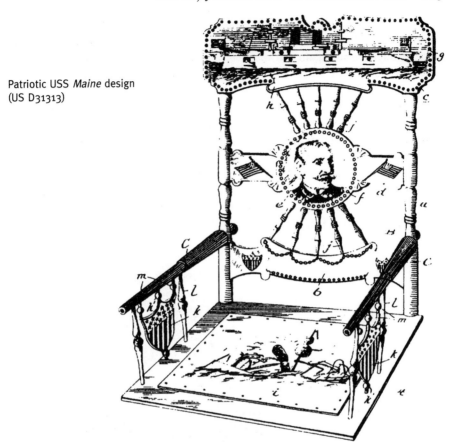

Patriotic USS *Maine* design
(US D31313)

sabotaged the ship. An American enquiry quickly announced that sabotage had been proven and, to the cry of "Remember the Maine !" war was declared. After a relatively quick and bloodless war the Spanish surrendered Guam, Puerto Rico and the Philippines to the USA, and Cuba achieved its independence.

The wonderful design shown was applied for in July 1899 and became US D31313. It is by Charles Kaiser of Rockport, IN. It shows the USS *Maine* in two views, one untouched and the other when it had been blown up. The inventor probably did not think of the disrespect that would come from the user's posterior resting on the sunken ship. The President at the time, McKinley, is also shown.

A US official enquiry in 1978 established that the actual reason for the explosion was a buildup of gases in the coal bunker, as established by examining the pattern of damage—although this itself has been criticized as faulty reasoning. Sometimes mistakes lead to turning points in history. A number of other patriotic designs were applied for during the war, including a bizarre "box" by Alfred Faxon of Philadelphia with US D29057. The box is covered with Old Glory while a depiction of a grinning Uncle Sam is on a board sticking up at the back. Its exact use is not clear. Uncle Sam also turns up as a kite (US D28905) but it is the USS *Maine* which appears again and again: as a smoker's pipe (US D28787), as an intricately designed plate with flag and eagle (US D28840), and on a carpet (US D29164), as well as various warship toy banks.

Similarly, from April 1917 World War I led to many patriotic design patents for such objects as playing cards, combs, finger rings, pins, car radiators, easels and lampshades, with photograph holders perhaps being the favorite. The exact depictions varied wildly, with perhaps "USA" between American flags being the most common, and great ingenuity shown in variations on existing designs. Over 50 designs used Old Glory. Uncle Sam again features as the King in a playing card suit (US D51574) and also, combined with the eagle motif, as a splendid statue in US D51415 by Mary Harris of San Francisco.

An unusual Uncle Sam (US D51415)

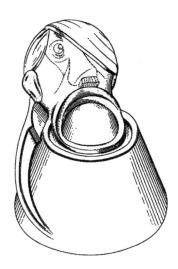

Patriotic designs
from World War II.
Clockwise from top
left, US D132186,
D134162, D135118,
D135430, D134824

Then there is US D52871 by William Flint of Los Angeles which shows a child wrapped in the flag apparently stamping on the Kaiser, with "Kaiser's Finish" at the bottom of the statue. Similarly Graham Trippe of Atlanta with US 1282149 had a board game for a race to Berlin which he called "Kop the Kaiser", while Antonin Lasalle of San Francisco's US 1350653, "Target", is for a mechanism where an American soldier strikes at the head of a bemused Kaiser. Nevertheless, most inventive activity of the time had nothing to do with the war, while the existence of a design patent did not mean that anything of the kind was necessarily ever manufactured.

Local loyalty occurred as well, as in D121385, filed in 1940 by a couple of Houston residents. This was a badge which placed on an equal footing the ideals of Americanism, loyalty, liberty, justice... and Texas. From 1942 there came another big surge in patriotic design patent applications before war weariness reduced activity. This time they were mainly for flags or pins. Many celebrated America while others commemorated links with allies such as Britain, China or Russia—or the United Nations as such, even before it was formed (US D135364, in 1942), as that was the name often used to describe the Allies. A small number were politically incorrect, such as cuspidors featuring the mouths of Hitler (US D134162) and Mussolini (US D134163) both filed by Edward Grossman of Chicago in 1942. The Mussolini design contrasted strongly with the sentiments of John Inquisitore's US D121216 in 1939 for a medal showing the Duce and the "fascines" (sticks) symbolic of his regime.

9/11 Illuminated memorial device (US D475500)

The terrible cost of war was also something to consider. Bessie Blount, a physical therapist, worked in veterans' hospitals helping amputees use their feet in place of hands. She thought of a way in which armless people could feed themselves and applied in 1948 for US 2550554. The device is worn around the neck and biting on a tube causes food to come down it. The Veterans Administration declined to use her idea, but the French government took it up. She later said that she hoped that she had proved "that a black woman can invent something for the benefit of humankind". Over 30 patents have been taken out by the Administration since then, such as US 4030141, "Multi-function control system for an artificial upper-extremity prosthesis for above-elbow amputees", in 1976.

The vogue for registering patriotic designs seems to have faded after World War II with only a few in the Korean War, and no apparent surge in anti-Communist emblems in the Cold War era. Some board games were however patented such as US 5020805 in 1989 where the players try to free POWs held in Vietnam, and two games based on the 1991 Gulf War, US 5108112 and US 5465973. The recent surge in terrorism may lead to more patriotic designs, such as Cindy Zou's 2003 design patent illustrated opposite.

There will always be arguments about the nature of the American Dream and whether it is always desirable. One thing at least is certain: if freedom is part of the American Dream, then it is worth making sacrifices for.

Further reading

··

THERE are many sources for this book, but the chief one is utility patents, design patents and trademarks. With the exception of one British patent every single drawing in this book is taken from the publications of the US Patent and Trademark Office (USPTO). All 6 million plus American utility and design patents are available to see or copy free of charge from the web, and that is why this book cites so many patent numbers. An explanation of how to look them up, or to carry out research on the web, is given below, but it is always best to begin research in a patent library.

Inspiration for the book came from a variety of sources. There was a great deal of material available to me that did not "fit" in my earlier *Inventing the 20th century: 100 inventions that shaped the world* (2000) and *Inventing the 19th century: the great age of Victorian inventions* (2001) also published by The British Library and New York University Press. Overlap with those books was deliberately avoided. Much searching was also carried out on the web to see what others thought significant and interesting, as virtually every topic, even the apparently most mundane, seems to have at least one devotee. Many of them have placed information on the web, and I am grateful for their enthusiasm and knowledge. Nevertheless much work involved the old-fashioned method of looking things up in printed sources such as annual indexes and plowing through material in the USPTO's *Official Gazette*. There are plenty of other topics that can be researched which are not covered, or which are only touched on, in this book, and I would be delighted if anyone feels stirred to take up the challenge. Trademarks are only briefly mentioned and deserve a book to themselves. (The British Library will be publishing *British Trademarks: a history* by David Newton in 2004.)

Some of these websites mention patents, but few give actual patent numbers, and fewer still provide scanned images from the patents on their sites. This is a shame, as no copyright is held in the original images, and many are very attractive. A picture is indeed often worth a thousand words. Surfing on a search engine like Google at **http://www.google.com** will reveal much: it is simply a matter of typing in variations on, say, ⟨baseball patent⟩ or ⟨"hanger was invented"⟩ to see if anything looks interesting. Books, of course, were also valuable. Not everyone or everything could be included, and some classics of American popular culture (such as Marilyn Monroe and Elvis Presley) could not, as far as I could tell, be represented by utility or design patents. A real problem with research is that the titles of design patents often merely say, e.g., "Toy" rather than "R2D2 robot". A problem with using sources on the web is that many of them gave different accounts of the origin of inventions, and it is hard to decide which are urban legends. That is what is so valuable about the published patents and other materials from the USPTO: they are "hard" facts which provide a firm foundation for research.

This chapter briefly explains how to go about researching the history of particular topics, or to look for companies or inventors, or to find specific patents. A simple way to see how a concept has evolved is to find a recent one on the subject and to see if it discusses earlier patents on the same subject. Many do, normally pointing out problems with the earlier patents which they, of course, will solve. This is a very useful approach for schools or those teaching design: show them the earlier patent and ask them to guess what problems it has and then show them the later patent. If you want to go into detail in research there is, however, only so much that you can learn by accessing the web, and not all sources are available there anyway. Spelling mistakes and omissions have also been known on databases (as in printed sources), and they do not always index what you want or the way you want, so caution is always warranted: it is practically impossible to find everything on a subject no matter how carefully you work. It is always best to start by visiting your closest patent collection and asking for help.

Researching in patent libraries

There is much to be said in carrying out research, at least initially, at a patent collection even if what you want is on a free database on the web. It is often difficult to make a correct or useful search without guidance. Anyone thinking of patenting an invention should definitely consult a patent collection first.

The United States Patent and Trademark Office (USPTO) has a huge library at its headquarters in Arlington, VA. This Scientific and Technical Information Library can be telephoned on (703) 308-1076. The USPTO also supports over 90 Patent and Trademark Depository Libraries across the USA. They are mostly located in public libraries, state libraries or land grant universities and can be visited to carry out research. It is always best to telephone before a visit. They are listed at **http://www.uspto.gov/go/ptdl/ptdlib_1.html**. Holdings will vary, but will often include a long run of the *Official Gazette* (which gives drawings and claims for utility patents from 1872, and for design patents from 1892). Annual name and subject indexes may also be held.

For those living in Europe, similar networks operate in some European countries. These are listed at **http://www.bl.uk/services/information/patents/othlink2.html#lib**.

Researching databases on the web

The British Library has a detailed site at **http://www.bl.uk/patents** for those who wish to research by themselves. This includes numerous links to databases across the world, often with hints on how to use them, and advice on searching techniques. The **http://www.bl.uk/services/information/patents/history.html** page is designed for those who wish to identify something with patent numbers on it.

There are three main databases, all free, which hold all American utility patents. Two of them also hold the design patents. They can be searched by the actual number, but other search elements will vary. The patents can be freely copied and reproduced as no copyright is claimed in them. The sites are all listed with brief comments at http://www.bl.uk/services/information/patents/keylinks.html. The explanations for each database given below date from March 2003, and facts may have altered since. They may seem difficult to use. If so it is again best to use a patent library.

The **DEPATISnet** database at http://www.depatisnet.de has been created by the German Patent Office. It has both utility and design patents. It is best used to see a known publication number. Select the British flag icon to select an English-language search page and then the "Beginner's search" option. In the top, "publication number" box select US for the first box and enter the number in the second box. This number can be a design or a utility patent number. The third box can be ignored. Click on "Start search". Normally two options are presented, each with a "PDF" icon. The number ending with A is a utility patent, that with S a design patent. Click on the icon to view an Adobe Acrobat copy of the first page. You can navigate within it and improve the resolution if you wish.

The **Esp@cenet** database at http://gb.espacenet.com has been created by the European Patent Office. It only has utility patents (but some design patents have recently been added). Select the red "Worldwide" option. To see a utility patent number enter e.g. US2861806 in the publication number box. Then click on the patent number to see the list of hits, a brief record and then the first page as an Adobe Acrobat image. You can navigate through the patent but not improve the resolution. To return to the brief record click on "Back" and then on the binoculars icon to return to the search page or on the icon below it to return to the list of hits. Most records for US patents from 1920 to the present will also have "EC" classifications as links. These can be looked at to see the definition of the inventive step involved. That for the Disney patent cited above is A63G31/16. The definition is that it is an amusement arrangement creating illusions of travel. This can be used in the search page in the "EC" search box, perhaps in conjunction with US in the "publication number" box (to limit the search to US patents). This kind of research can be fascinating and was very helpful in gathering material for this book.

Each search box can be filled in to add more concepts to the search. For example, filling in "Disney" in the applicant box limits the search to Disney as a person or company assignee (and gives over 300 hits). By putting say the word "amusement" in the title box at the same time, the results are limited to a few dozen. Adding an * to the end of the word truncates it so that child* means child, children or any other word beginning with child. Again, endless pleasure can be obtained by this searching. However, there are believed to be gaps, and other than the patent number these search elements are limited to 1920 onwards.

The official **USPTO database** at http://www.uspto.gov/patft/index.html indexes by classification back to the beginning for both utility and design patents. However, search elements other than the actual number are only available from 1976. There are two databases: for grants (including design patents), on the left of the home page, and for applications published since March 2001 on the right. The publication number option explains how to search for a known number. In order to see the actual patent software must be loaded from the "Help" key (this is free). The British Library itself uses the AlternaTIFF option.

The other two options are quick and advanced search pages. Quick is adequate for most purposes. There are two boxes to fill in with words or numbers plus boxes next to them where search elements can be selected. If they are not selected, then every single patent back to 1976 is searched—over 2 million of them. This is best used for unusual words. Alternatively say title, abstract and so on can be selected. US classifications can also be used in the format 273/316.6. The search always defaults to the last few years and this can be extended to "all years" by clicking on the "select years" box. Truncation is possible but with a $ sign.

The search is still limited to 1976 onwards except for searches by class. The classes can be guessed at by using the results from other searches but are best checked in the "Tools to help in searching by patent classification" page which is listed on the home page for the databases.

Searching by name can cause problems. On the **Esp@cenet** database there are no entries for Walt Disney, but four for Walter. Bobby may be Robert, and so on. Some abbreviations are used for corporate bodies, such as univ for university, gov for government, and inst for institute. Again, patent libraries are used to such problems and can usually help.

Websites about inventions

For those who wish to learn about particular inventions or subject areas the Inventors.About site at **http://inventors.about.com** is invaluable. For those who wish to understand how machines actually work, Marshall Brain's HowStuffWorks is similarly packed with fascinating information on how numerous inventions work at **http://www.howstuffworks.com**. If funny patents are of interest then the British Library lists a number of websites with drawings from funny patents at its **http://www.bl.uk/services/information/patents/othlink3.html#fun site**.

There are many other useful sites including:

Index of African American Inventors: Historical
[By name giving inventions and, sometimes, date and/or patent number.]
http://www.princeton.ecu/~mcbrown/display/inventor_list.html

National Inventors Hall of Fame Index of Inventions
[By invention giving patent number and information on inventor]
http://www.invent.org/hall_of_fame/1_1_search.asp

Rothschild Petersen Patent Model Museum
[The largest private collection of models of what the old inventions looked like]
http://www.patentmodel.org/index.html

Thomas Edison's patents
[Lists and links to copies of all 1,093 patents]
http://edison.rutgers.edu/patents.htm
http://www.shouldexist.org and **http://www.halfbakery.com** are also very useful.

Material on the patent system

For those interested in the patent system generally, Edward Walterscheid's *To promote the progress of useful arts: American patent law and administration, 1798–1836* (1998) and Kenneth Dobyns' *The Patent Office pony* (1994), which is more light-hearted, are excellent for the first half of the US Patent and Trademark Office's history. **http://www.myoutbox.net/polist.htm** contains the annual reports for 1837 to 1892 plus much other material relating to that period. No authoritative account of the second half of the USPTO's history has apparently been published.

The USPTO annual reports are rich in interesting comments and give data for seeing how many patents each state had, and showing which were the most productive. Recent annual reports, and a vast amount of statistics analysing, for example, companies, technologies, counties and metropolitan areas, are available at **http://www.uspto.gov/web/offices/ac/ido/oeip/taf/index.html**. For modern material there is plenty on **http://www.uspto.gov**.

Books about inventions

This is a selection, concentrating on those with many patent drawings.

Absolutely mad inventions. A. E. Brown, H. A. Jeffcott Jr, 1932, reprinted 1970.

Inventions necessity is not the mother of: patents ridiculous and sublime. S. V. Jones, 1973.

American sex machines: the hidden history of sex at the U.S. Patent Office. H. Lewis, 1996.

Banana bats and ding-dong balls: a century of unique baseball inventions. D. Gutman, 1995.

Historical first patents: the first United States patent for many everyday things. T. Brown, 1994.

Mr McMurtry's bubble hat . . . and other great moments in American ingenuity. M. Miller, 1996.

The practical inventor's handbook. O. Greene, F. Durr, 1979 [More fun than it sounds, with numerous attractive drawings].

Totally absurd inventions: America's goofiest patents. T. VanCleave. 2001.

Inventing the American fire engine: an illustrated history of patented ideas for fire pumpers. M. Goodman. 1994.

Professor Henry Petroski is one of the few historians of technology to make extensive use of patents in his writing about pencils, paperclips and other apparently mundane objects in, for example, *The evolution of useful things* (1993). Daniel Boorstin, a former Librarian of Congress, did not use patent drawings in his books which show a keen awareness of the impact of technology on American life, especially in his *The Americans: the democratic experience* (1973).

Asking for further help

The British Library has long taken an interest in historical patents from all countries. It holds 46 million patents from over 40 countries, arranged in number order. A great deal of experience has been built up by its staff in dealing with a wide range of enquiries on many different aspects of intellectual property, and it is doubtful if any other library is as experienced in dealing with historical or humanities-based enquiries to do with patents. Enquiries are welcome provided it is understood that only half an hour's free help is normally provided, and that legal advice is not provided. Visitors who wish to carry out research in person are also welcome but a reader's ticket is required (see **http://www.bl.uk/services/reading/admissions.html**). If contacting us from outside Britain it is best to fax or email rather than to write or to telephone.

Contact details are:

The British Library
Patents Information
96 Euston Road
London NW1 2DB
UK
Tel. 020 7412 7916 (from inside UK)
Fax 00 44 20 7412 7480 (from outside UK)
Email patents-information@bl.uk
Website **http://www.bl.uk/patents**

Index

. . . .

This index is selective for names and trademarks, and only indexes major cities within lengthy headings for each state.

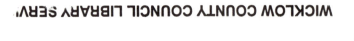